Introduction to Structure and Properties of Atoms

INTRODUCTION TO STRUCTURE AND PROPERTIES OF ATOMS

Serhii Solodin

ARCLER PRESS

www.arclerpress.com

Introduction to Structure and Properties of Atoms

Serhii Solodin

Arcler Press
224 Shoreacres Road
Burlington, ON L7L 2H2
Canada
www.arclerpress.com
Email: orders@arclereducation.com

ISBN: 978-1-77469-940-9 (Hardcover)

ABOUT THE AUTHOR

Serhii Solodin is a PhD in Chemistry (2021) and the author of 14 scientific publications rated in Scopus database. His thesis and scientific interest concerned the study of defects in semiconductor crystals, particularly Cadmium Telluride, which was carried out in Poland, the Czech Republic and Austria. Additionally, he spent the past years researching and developing products to solve consumer issues in the personal care and cosmetic industry. Among the products he has developed (340+) there are products for skin care (creams, serums, serums, gels), hair care (shampoos, conditioners, masks, sprays), oral care (toothpaste and tablets), laundry care (gels, tablets, sheets). He likes solving extraordinary, new and challenging issues and always strives for self-improvement and constantly monitor the market to keep abreast of current trends and approaches to product creation. In particular, recently he has been developing products aimed at environmental friendliness and biodegradability, in particular, products without plastic: effervescent tablets and powders for oral cavity, laundry, skin- and hair-care as well as detergent sheets.

TABLE OF CONTENTS

LIST OF FIGURES

LIST OF TABLES

LIST OF ABBREVIATIONS

AAS	Atomic Absorption Spectroscopy
AES	Atomic Emission Spectroscopy
AFM	Atomic Force Microscopy
CEAS	Cavity-Enhanced Absorption Spectroscopy
CRT	Cathode Ray Tube
CS AAS	Continuum Source Atomic Absorption Spectroscopy
CSIRO	Commonwealth Scientific and Industrial Research Organization
DFT	Density Functional Theory
EDLs	Electrodeless Discharge Lamps
ESI	Electrospray Ionization
ET AAS	Electrothermal Atomic Absorption Spectroscopy
FIB	Focused Ion Beam
HCLs	Hollow Cathode Lamps
ISAS	Institute of Spectrochemistry and Applied Spectroscopy
LCAO	Linear Combination of Atomic Orbitals
LIBS	Laser-Induced Breakdown Spectroscopy
LS AAS	Line Source Atomic Absorption Spectroscopy
MALDI	Matrix-Assisted Laser Desorption/Ionization
MOs	Molecular Orbitals
NMR	Nuclear Magnetic Resonance
PET	Positron Emission Tomography
QFT	Quantum Field Theory
RF	Radio Frequency
STM	Scanning Tunneling Microscopy
STPF	Stabilized Temperature Platform Furnace
TEM	Transmission Electron Microscopy
TOF-MS	Time-of-Flight Mass Spectrometry

PREFACE

This book examines the fundamental nature of atoms, their intricate structures, and the multitude of properties resulting from their interactions. It serves as a comprehensive guide to comprehending the essence of atoms and their role in shaping the physical and chemical landscapes of our planet. As the exploration of atomic theory progresses, insight is gained into the evolution of human thought and scientific inquiry, tracing the path from ancient philosophical musings to contemporary chemistry's cutting-edge insights.

Chapter 1 establishes the foundation by tracing the historical origins of atomic theory and introducing the fundamental concepts that define an atom. The early speculations of Greek philosophers, groundbreaking experiments of scientific pioneers, and pivotal discoveries that underpin our contemporary understanding of atoms are investigated.

Chapter 2 illustrates the evolution of atomic models, from Dalton's indivisible spheres to the quantum mechanical formulations characterizing the behavior of subatomic particles. The transformation of our understanding of the atom is observed as models are constructed to explain its behavior, leading to the profound principles governing its conduct.

Chapter 3 explores the field of light and color emitted by atoms. It examines the intricate dance of electrons as they transition between energy levels, yielding mesmerizing spectra that reveal the unique characteristics of each atom. This chapter unveils the concealed secrets within light emission and its role in disclosing the mysteries of atomic structure.

Chapter 4 goes beyond individual atoms to explore the forces binding them together to create molecules. The complex mechanisms regulating chemical reactions and shaping the diverse array of substances encountered in daily life are examined.

Chapter 5 scrutinizes the apparent simplicity of atomic dimensions, exploring the subtleties of atomic radii and explaining how these seemingly inconspicuous dimensions exert influence over a wide range of physical and chemical properties, from the compact structure of solids to the interactions among molecules.

Chapter 6 introduces the realm of electrons, where the principles of quantum mechanics hold preeminent sway. It unveils the intricate arrangement of electrons within atoms, determining their behavior and giving rise to the variety of elements in the periodic table.

Chapter 7 bridges the gap between theory and application by demonstrating how the fundamental understanding of atoms propels contemporary advancements in materials science, nanotechnology, and related fields. It illustrates how our comprehension of atomic structure empowers the creation of new substances and manipulation of matter at the atomic level.

Chapter 8 sheds light on the state-of-the-art techniques and technologies that enable the examination of atomic behavior with unprecedented clarity. The domains of spectroscopy, microscopy, and computational modeling are explored, allowing the revelation of intricate details of atomic behavior and the design of novel materials for a rapidly evolving world.

CHAPTER 1

Fundamentals of Atom

Contents

1.1. INTRODUCTION

The study of atoms, which are the fundamental constituents of matter, has been an essential domain throughout the annals of scientific research. This journey through the microscopic realm of atoms, spanning from ancient philosophical speculations to the revolutionary advancements in modern physics, has significantly influenced our comprehension of the fundamental nature of the universe.

The notion of atoms has captivated the human intellect for countless centuries. The concept of indivisible particles, known as "atoms," considered the immutable essence of all matter, was postulated by the ancient Greek philosopher Democritus. The aforementioned philosophical speculation gained empirical support in subsequent centuries through pioneering scientists who attempted to unravel the mysteries surrounding atoms and their significance in chemical interactions.

This chapter illustrates the foundational aspects of atoms, which are essential for comprehending the essence of matter and the universe as a whole (Gao et al., 2021). Atoms, the fundamental building blocks of matter, contain smaller elementary particles, leading to their complexity and our understanding of the natural world.

Within the structure of every atom, protons, neutrons, and electrons exist. Protons carry a positive electric charge and reside in the atomic nucleus, while neutrons are electrically neutral and also inhabit the nucleus. Electrons, with their negative charge, orbit the nucleus within discrete energy levels. These particles, despite their small size, play a crucial role in determining the characteristics and behavior of atoms (Coessens & Matyjaszewski, 2010).

Two fundamental properties of atoms include the atomic number, indicating the number of protons and determining the element, and the mass number, representing the combined total of protons and neutrons in the nucleus. Isotopes, different forms of an element with the same atomic number but varying mass numbers due to different neutron quantities, have essential applications, from dating archaeological artifacts to medical imaging using radioactive isotopes (Matyjaszewski & Xia, 2002).

Niels Bohr's atomic spectrum theory, proposed in 1913, explains that electrons have discrete energy levels while orbiting the nucleus, leading to the emission of distinct spectral lines during transitions between these levels. This theory advanced our understanding of atomic structure and the phenomena of light emission and absorption by atoms (Wang et al., 2021).

As the chapter progresses, the wave-like characteristics of electrons come to the forefront, challenging the conventional view of particles as localized entities. The concept of wave-particle duality, stemming from quantum mechanics, reveals that electrons exhibit characteristics of both particles and waves, revolutionizing our understanding of the atomic world (Cui et al., 2022).

Werner Heisenberg's Uncertainty Principle is a concept of great interest in atomic physics. It explains that knowing both the momentum and precise position of a subatomic particle simultaneously is unattainable. This principle has profound implications for our understanding of reality within the realm of quantum mechanics, sparking fascination in both scientific and philosophical communities (Nguyen et al., 2020).

The Schrödinger Wave Equation, originally proposed by Erwin Schrödinger, serves as a fundamental equation in quantum mechanics. It provides a framework for understanding the wave-like nature of electrons and their behaviors. This equation plays a vital role in explaining atomic orbitals, which are spatial regions with a high probability of electron presence. Atomic orbitals have distinct geometries and energy levels, forming the basis for the organization of electrons within atoms (Khandelwal et al., 2022).

The chapter further explores the complexities of atomic structure and the organization of electrons within energy levels and orbitals. The chemical properties of an atom are determined by the interactions between its electrons and the nucleus, forming the foundation for the diverse range of elements and their behaviors (Loncar, 2021). The periodic table, a cornerstone of chemistry, organizes elements based on their atomic number, facilitating the identification of patterns in reactivity, atomic size, and other properties. It offers a structured framework for understanding the wide variety of elements and their compounds (Abdelghafar et al., 2022).

Lastly, the chapter looks into nuclear chemistry, focusing on nuclear reactions, radioactivity, and the substantial energy potential associated with nuclear fission and fusion processes. Nuclear chemistry has far-reaching implications in energy production, medical applications, and scientific research, making it a crucial field of study in our modern society (Zhuang et al., 2020; Li et al., 2022).

Tip

The size of atoms is such that the quantity of atoms present in a single droplet of water surpasses the number of stars observed within the universe.

1.2. HISTORICAL DEVELOPMENTS

The concept of matter composed of discrete units is an ancient notion traced back to various civilizations, such as Greece and India. The term "atom" (derived from the Greek word "atoms"), which signifies "indivisible," was introduced by the ancient Greek philosophers Leucippus and his disciple Democritus (c. 460–c. 370 BC) of the Pre-Socratic era. Democritus espoused the belief that atoms possess infinite quantity, are devoid of a beginning or end, and endure indefinitely. Furthermore, he posited that the attributes exhibited by an object are a direct consequence of the specific atoms that constitute it. The concept of atomism, initially proposed by Democritus, underwent further development and expansion by subsequent Greek philosopher Epicurus (341–270 BC) and Roman Epicurean poet Lucretius (c.99–c.55 BC). Atomism was largely neglected in Western Europe during the Early Middle Ages. In the 12th century, there was a resurgence of awareness regarding Aristotle's teachings in Western Europe, leading to its renewed recognition in this region. Aristotle's contrasting perspective on matter posited its continuity and infinitude, allowing for limitless subdivision (Viana & Porto, 2010).

During the 14th century, there was a resurgence of interest in atomist teachings due to the rediscovery of significant ancient works such as Lucretius's De rerum natura and Diogenes Laërtius's Lives and Opinions of Eminent Philosophers. This led to a notable increase in scholarly focus on the subject matter. However, due to its association with the philosophical school of Epicureanism, which conflicted with orthodox Christian doctrines, the acceptance of atomism was largely deemed unacceptable among most European philosophers (Weinert, 2000). Pierre Gassendi, a prominent French Catholic priest during the late 16th and early 17th centuries, revitalized the philosophical concept of Epicurean atomism by introducing certain adaptations. Gassendi posited that atoms, which he believed were brought into existence by a divine entity, were indeed highly abundant but not boundlessly infinite in quantity. He was the inaugural proponent of the term "molecule" as a descriptor for the conglomeration of atoms. The modified theory of atoms proposed by Gassendi gained popularity in France through the efforts of the physician François Bernier (1620–1688) and in England through the contributions of the natural philosopher Walter Charleton (1619–1707). The proponents of atomism, namely chemist Robert Boyle (1627–1691) and physicist Isaac Newton (1642–1727), successfully advocated for the concept during the late 17th century. Consequently, a significant segment of the scientific community embraced the notion of an atomistic basis for the natural world (Rodríguez & Niaz, 2002).

1.3. FUNDAMENTAL PARTICLES OF AN ATOM

The mass of a neutron and a proton is approximately equal, while the mass of an electron is considered negligible in comparison to these particles. The proton possesses a positive charge that is equal in magnitude, yet opposite in sign, to the negative charge carried by an electron. Conversely, a neutron does not possess any charge. In the atomic structure of any given element, the quantity of protons is equivalent to the number of electrons, resulting in a state of electrical neutrality for the atom. The atomic nucleus comprises protons and neutrons and carries a positive charge. In the case of protium, the nucleus consists of a solitary proton.

The electrons are situated within a spatial region surrounding the nucleus. The majority of an atom's mass is localized within its nucleus, although the nucleus occupies a significantly smaller volume compared to the entire atom. Specifically, the radius of the nucleus is approximately 10^{-15} meters, while the overall size of the atom is roughly 10^5 times greater than this. It can be inferred that the nucleus density is significantly higher, surpassing the density of the metal Pb by a factor of more than 10^{13} (Charalampous, 2021).

1.4. ATOMIC NUMBER, MASS NUMBER, AND ISOTOPES

The identification of atoms is based on the quantification of protons and neutrons within their atomic structure. The term "atomic number" (Z) is the number of protons present within the nucleus of every individual atom belonging to a specific element. In the context of a neutral atom, it is observed that the quantity of protons corresponds to the quantity of electrons, thereby establishing a balance. Consequently, the atomic number serves as an indicator of the quantity of electrons that exist within the atom

(Oganessian et al., 2010). The determination of an atom's chemical identity can be exclusively achieved through its atomic number. As an illustration, it can be observed that the atomic number of nitrogens is 7, indicating that every nitrogen atom in a neutral state possesses 7 protons and 7 electrons. Alternatively, it can be argued that any atom within the universe possessing seven protons is accurately designated as "nitrogen." The mass number (A) represents the combined count of neutrons and protons within an element's atomic nucleus. With the exception of the most prevalent variant of hydrogen, characterized by a solitary proton and the absence of neutrons, all atomic nuclei consist of a combination of protons and neutrons (Kratz et al., 2015). Typically, the mass number is determined by:

$$mass\ number\ (A) = number\ of\ protons + number\ of\ neutrons$$
$$= atomic\ number\ (Z) + number\ of\ neutrons$$

The quantity of neutrons present within an atom can be determined by subtracting the atomic number (Z) from the mass number (A), expressed as (A–Z). As an illustration, in the case of a specific boron atom with a mass number of 12 and an atomic number of 5, the number of neutrons can be evaluated by subtracting the atomic number from the mass number, resulting in 7 neutrons. It is important to acknowledge that the atomic number, number of neutrons, and mass number are required to be positive integers, also known as whole numbers. In the majority of instances, atoms belonging to a specific element exhibit varying masses. Isotopes are defined as atoms with identical atomic numbers but varying mass numbers. As an illustration, it is noteworthy to mention that hydrogen exhibits three distinct isotopes. The element commonly referred to as hydrogen consists of a single proton and lacks any neutrons. The deuterium isotope is characterized by the presence of one proton and one neutron, while tritium is distinguished by its composition of one proton and two neutrons. The conventional method for representing the atomic number and mass number of an atom of element X is as follows (King & Fall, 1940):

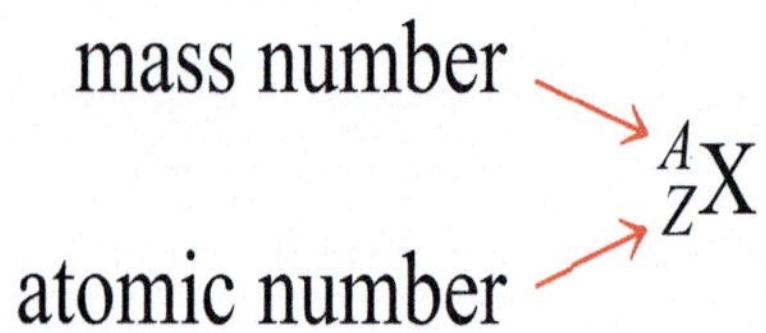

Thus, hydrogen isotopes,

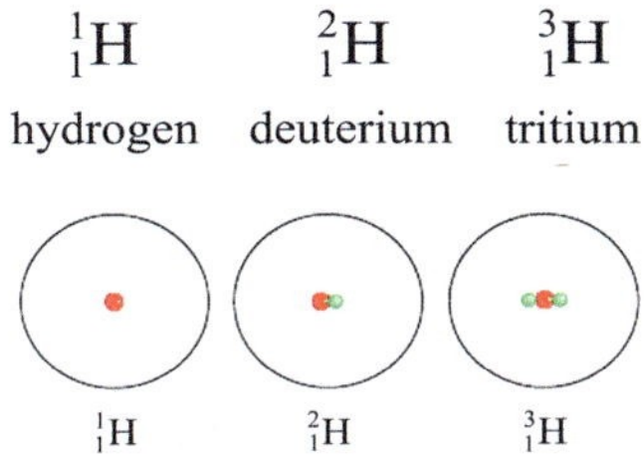

To further illustrate this point, let us examine two prevalent isotopes of uranium, characterized by their respective mass numbers of 235 and 238 (Alsaif et al., 2017):

$$^{235}_{92}U;\ ^{238}_{92}U$$

The initial isotope finds utility in nuclear reactors and atomic weaponry, while the latter isotope lacks the requisite characteristics for such applications. Except for hydrogen, identifying isotopes of elements is based on their respective mass numbers. Hence, the isotopes are referred to as uranium-235 and uranium-238. The chemical characteristics of an element are primarily dictated by the protons and electrons present within its atomic structure, while neutrons generally remain inert and do not actively participate in chemical reactions under typical circumstances. Consequently, isotopes belonging to the same chemical element exhibit analogous chemical properties, giving rise to identical compound formations and demonstrating comparable levels of reactivity (Kurudirek, 2017).

1.4.1. Nuclides, Atomic Number, and Mass Number

A nuclide refers to a specific category of atoms that exhibit a distinctive atomic number, denoted as Z, which corresponds to the quantity of protons present within the nucleus. It is worth noting that due to the atom's electrical neutrality, Z is also equivalent to the number of electrons. The mass number, denoted as A, represents the total number of neutrons and protons contained within the nucleus of a given nuclide (Block et al., 2010). The utilization of a concise technique to represent the atomic number, mass number, and symbol of a nuclide, denoted as E, is:

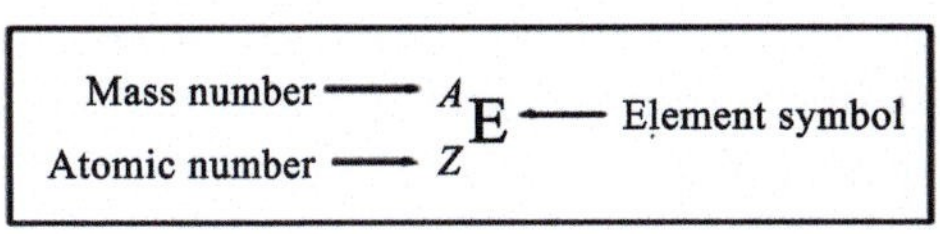

e.g. $^{20}_{10}Ne$

Atomic number = Z = number of protons in the nucleus = number of electrons

Mass number = A = number of protons + number of neutrons

Number of neutrons = A – Z

1.4.2. Relative Atomic Mass

The mass of an atom is primarily determined by the number of neutrons and protons in its nucleus, as electrons have negligible mass. The mass of an individual atom is infinitesimally small and typically non-integer. To facilitate calculations and comparisons, we employ a system of relative atomic masses.

The atomic mass unit is conventionally defined as one-twelfth of the mass of a carbon-12 atom, resulting in a value of approximately 1.660×10^{-27} kilograms. The relative atomic masses (Ar) are expressed in relation to the isotope ^{12}C, which is assigned a value of 12.0000.

The masses of the proton and neutron can be approximated to be approximately 1 atomic mass unit (u), where 1u (also known as Dalton) is approximately equal to 1.660×10^{-27} kilograms (Lodders, 2021).

1.4.3. Isotopes

Nuclides belonging to the same chemical element exhibit identical quantities of protons and electrons, yet they may exhibit distinct mass numbers. The element is determined by the number of protons and electrons it possesses, while the number of neutrons may exhibit variability. Isotopes are defined as nuclides of a specific element that exhibit variations in the number of neutrons, resulting in different mass numbers. Certain elements have naturally occurred isotopes, while others can be artificially produced (Peterson & Fry, 1987).

Monotopic elements, such as phosphorus (P) and fluorine (F), are characterized by their occurrence in nature with only one nuclide. Isotopic mixtures encompass certain elements, such as carbon (C), with isotopes: and , as well as oxygen (O) with isotopes .

Isotopes are frequently differentiated solely by indicating their atomic masses, such as ^{12}C and ^{13}C, due to the fact that the atomic number remains constant for a specific element (Firestone & Shirley, 1998) (Figure 1.1).

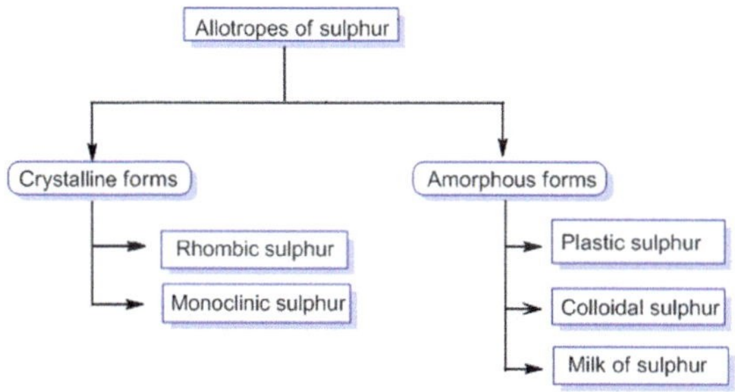

Figure 1.1. *Allotropes of sulfur.*

Source: Chemical Note, Creative Commons License.

1.4.4. Success in Early Quantum Theory

The electrons within an atom are situated within a spatial region surrounding the nucleus. The significance of electrons in dictating the characteristics of atoms, ions, and molecules, as well as their bonding interactions, necessitates a comprehensive detail of the electronic configurations of each entity. A comprehensive examination of electronic structure necessitates the inclusion of quantum theory and wave mechanics. The evolution of quantum theory occurred in two distinct phases. The electron was regarded as a discrete particle in the earlier theoretical frameworks spanning from 1900 to 1925. Notably, during this period, the most significant contributions to the field of inorganic chemistry were the explanation of atomic spectra and the determination of electronic configurations. In contemporary models, the electron is regarded as a wave, leading to the nomenclature of wave mechanics. Notably, significant advancements in the field of chemistry have been made, particularly in the understanding of stereochemistry fundamentals and the development of molecular property calculation techniques. However, it is important to note that these calculations are only precise for species containing light atoms. Given that wave mechanics can yield the same outcomes as the older quantum theory, it may appear redundant to refer to the latter. In fact, advanced treatments of theoretical chemistry rarely do so (Gearhart, 2010).

1.4.5. Some Important Success of Classical Quantum Theory

The examination of the historical advancements in quantum theory has been extensively covered in other scholarly works. Therefore, our attention will be directed solely toward certain fundamental aspects of classical quantum theory, wherein the electron is regarded as a particle. At lower temperatures, the emission of radiation from a heated object predominantly consists of low-energy infrared radiation. However, as the temperature rises, the emitted radiation transitions through a sequence of colors, from a dull red to a brighter red, ultimately culminating in a white hue. Efforts to explain this particular observation proved unsuccessful until 1901, when Planck proposed the concept that energy could solely be absorbed or emitted in discrete units known as quanta. These quanta are directly proportional to the frequency of the radiation, denoted as υ, and can be mathematically expressed by Eqn. (1.1). The constant of proportionality is denoted as h, which represents the Planck constant ($h = 6.626 \times 10^{-34}$ J s) (Dini & Hammer, 2017).

$$\Delta E = hv \ Units{:}\ E\ in\ J;\ v\ in\ s^{-1}\ or\ Hz \tag{1.1}$$

$$c = \lambda v \ Units{:}\ \lambda\ in\ m;\ v\ in\ s^{-1}\ or\ Hz \tag{1.2}$$

The hertz, denoted as Hz, is the internationally recognized unit of measurement for frequency in the International System of Units (SI). The relationship between the frequency of the radiation and its wavelength λ can be expressed using Eqn. (1.1), where c represents the speed of light in a vacuum. By rearranging Eqn. (1.1), we can derive Eqn. (1.3), which allows us to establish a connection between the energy of radiation and its wavelength (Glauber, 1963):

$$\Delta E = \frac{hc}{\lambda} \tag{1.3}$$

Based on this correlation, Planck derived a relationship between relative intensity, wavelength, and temperature that exhibited strong concordance with empirical evidence. The derivation process is complex and will not be presented in this context. The Rutherford-Bohr model of the atom played a crucial role in the early quantum theory by providing a framework for interpreting the atomic spectrum of hydrogen, which is considered to be one of its significant applications. When subjecting a dihydrogen sample to an electric discharge, the H_2 molecules undergo dissociation into individual atoms (Woolley, 1998). Consequently, an electron within a specific excited H atom has the potential to transition to various higher energy levels. The states above exhibit transience, wherein the electron undergoes a transition to a lower energy state, subsequently releasing energy in the process. The outcome is the detection of spectral lines in the emission spectrum of hydrogen. In the year 1885, Balmer made an observation regarding the wavelengths of spectral lines observed within the visible range of the atomic spectrum of hydrogen. Balmer noted that these wavelengths conformed to Eqn. (1.4), wherein R represents the Rydberg constant specific to hydrogen, υ denotes the wavenumber measured in cm^{-1}, and n represents an integer value of 3, 4, 5, and so forth. The collection of spectral lines that is commonly referred to as the Balmer series is widely recognized in the field (Dorai et al., 2018).

$$v = \frac{1}{\lambda} = R\left(\frac{1}{2^2} - \frac{1}{n^2}\right) \quad (1.4)$$

$$R = Rydberg\ constant\ for\ hydrogem = 1.097 \times 10^7 m^{-1} = 1.097 \times 10^5 cm^{-1}$$

Additional sets of spectral lines can be observed in the ultraviolet region, known as the Lyman series, as well as in the infrared region, specifically the Paschen, Brackett, and Pfund series. The general expression provided in Eqn. (1.5) is applicable to all lines in all series, with the condition that m is greater than n. The Lyman series corresponds to the principal quantum number n = 1, while the Balmer series corresponds to n = 2. Similarly, the Paschen, Brackett, and Pfund series correspond to n = 3, 4, and 5, respectively (Kochen & Specker, 1990) (Figure 1.2).

$$v = \frac{1}{\lambda} = R\left(\frac{1}{m^2} - \frac{1}{n^2}\right) \quad (1.5)$$

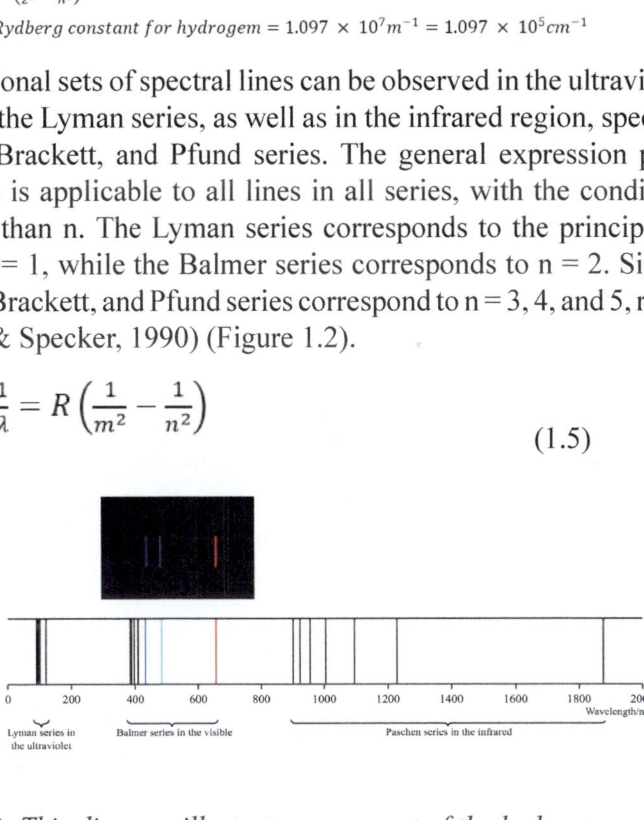

Figure 1.2. *This diagram illustrates a segment of the hydrogen spectrum, specifically highlighting the Lyman, Balmer, and Paschen series of emission lines.*

Source: Stephan, Creative Commons License.

Remember

The majority of an atom's volume is comprised of empty space, wherein electrons exhibit rapid motion as they orbit the nucleus.

1.5. BOHR'S THEORY OF THE ATOMIC SPECTRUM OF HYDROGEN

In the year 1913, Niels Bohr integrated principles from quantum theory and classical physics to develop a theoretical framework for analyzing the behavior of the hydrogen atom. He presented a pair of postulates concerning the behavior of an electron within an atomic system:

- Stationary states refer to specific energy levels in which the electron's energy remains constant. The mentioned states are characterized by their circular orbits around the nucleus, in which the electron exhibits an angular momentum represented by the expression mvr, as stated in Eqn. (1.6). In this particular context, the variable "n" is used to denote the principal quantum number (Benczik et al., 2005).

$$mvr = n\left(\frac{h}{2\pi}\right) \tag{1.6}$$

where; 'm' is the mass of electron; 'v' is the velocity of electron; 'r' is the radius of the orbit; 'h' is the Planck constant; and h/2 may be written as h.

- The absorption or emission of energy occurs exclusively during the transition of an electron between two stationary states. The energy change linked with this transition can be determined using Eqn. (1.7), where n1 and n2 represent the principal quantum numbers corresponding to the energy levels E_{n1} and E_{n2}, respectively (Bohr, 1913).

$$\Delta E = E_{n2} - E_{n1} = h\nu \tag{1.7}$$

The application of the Bohr model to the hydrogen atom allows for the determination of the radius of each permissible circular orbit through the utilization of Eqn. (1.8). The etymology of this phrase can be traced back to the centrifugal force exerted on the electron during its circular trajectory (Mills, 2000).

$$r_n = \varepsilon_0 h^2 n^2 / \pi m_e e^2 \tag{1.8}$$

where; ε_0 is the permittivity of a vacuum = $8.854 \times 10^{-12} Fm^{-1}$;

h is the planck constant = $6.626\ x\ 10^{-34} Js$;

$n = 1, 2, 3, \ldots$ describing a given orbit;

m_e is the electron rest mass = $9.109\ x\ 10^{-31} kg$;

e is the charge on an electron (elementary charge) = $1.602\ x\ 10^{-19} C$.

By substituting the value of n as 1 into Eqn. (1.8), it is determined that the radius of the first orbit of the hydrogen atom is 5.293×10^{-11} m, which is equivalent to 52.93 picometers (pm).

The quantity denoted by the symbol a0 is referred to as the Bohr radius of the hydrogen atom. The transition from a principal quantum number of n = 1 to n = ∞ holds particular significance. The phenomenon is the ionization of the atom, as represented by Eqn. (1.9).

The ionization energy, denoted as IE, can be determined by combining Eqns. (1.5) and (1.7), as demonstrated in Eqn. (1.10). The values of ionization energies (IEs) are expressed in terms of moles of atoms (Stetsko & Tkachuk, 2006).

$$H(g) \rightarrow H^{+}(g) + e^{-} \quad (1.9)$$

$$IE = E_{\infty} - E_1 = \frac{hc}{\lambda} = hcR\left(\frac{1}{1^2} - \frac{1}{\infty^2}\right)$$

$$= 2.179\,x\,10^{-18} J$$

$$2.179\,x\,10^{-18} x\,6.022\,x\,10^{23} J\,mol^{-1}$$

$$= 1.312\,x\,10^{6} J\,mol^{-1}$$

$$= 1312 kJ\,mol^{-1} \quad (1.10)$$

While the joule is the standard unit of energy in the International System of Units (SI), electron volts (eV) are commonly employed to denote ionization energies.

It is worth noting that 1 eV equals 96.4853 kilojoules per mole. Hence, the ionization energy of hydrogen may also be expressed as 13.60 electron volts (eV). The Bohr model demonstrated remarkable success in its application to the hydrogen atom. However, the inclusion of species with multiple electrons necessitated significant modifications to the model (Moumni et al., 2010).

1.6. AN INTRODUCTION TO WAVE MECHANICS

1.6.1. The Wave Nature of Electrons

The quantum theory of radiation, as postulated by Max Planck and Albert Einstein, posits the existence of particles associated with light, thus supplementing the wave theory of light that is necessary to explain phenomena such as interference and diffraction. Louis de Broglie proposed in 1924 that if light exhibited both particle and wave characteristics, then electrons and other particles should also possess a similar dual nature. The occurrence being discussed is commonly known as wave-particle duality. The de Broglie relationship, as expressed by Eqn. (1.11), integrates the principles of classical mechanics with the notion of wave-like characteristics. It demonstrates that a particle with momentum mv, where m represents the mass and v denotes the velocity of the particle, exhibits a corresponding wave with a wavelength λ (Roy, 2016).

$$\lambda = \frac{h}{mv} \text{ where h is the Planck constant} \tag{1.11}$$

1.6.2. The Uncertainty Principle

The presence of wave-like characteristics in electrons gives rise to a significant and challenging implication: the simultaneous determination of an electron's speed and location at a given moment becomes unattainable. The following is an explication of Heisenberg's uncertainty principle. To circumvent this issue, instead of attempting to determine the electron's position and momentum precisely, we employ the concept of probability to ascertain the likelihood of locating the electron within a specific spatial region.

1.6.3. The Schrodinger Wave Equation

The wavefunction is derived from the Schrödinger wave equation, which can be solved exactly or through approximate methods to obtain relevant information. The exact solution to the Schrödinger equation is only feasible for a species consisting of a nucleus and a single electron, commonly called a hydrogen-like system.

The representation of the Schrödinger wave equation can take various forms. In this analysis, we focus on its application to the movement of a particle confined within a one-dimensional box. Eqn. (1.12) provides the specific form of the Schrödinger wave equation that is suitable for describing motion along the x-axis (Salvat et al., 1995).

$$\frac{d^2\psi}{dx^2} + \frac{8\pi^2 m}{h^2}(E - V)\psi = 0 \tag{1.12}$$

where; 'm' is the mass; 'E' is the total energy; and 'V' is the potential energy of the particle.

$$\frac{\partial^2\psi}{\partial x^2} + \frac{\partial^2\psi}{\partial y^2} + \frac{\partial^2\psi}{\partial z^2} + \frac{8\pi^2 m}{h^2}(E - v)\psi = 0 \tag{1.13}$$

The resolution of this equation is not of immediate relevance, although it is worth mentioning that employing spherical polar coordinates can offer certain advantages. The analysis of outcomes derived from the Schrödinger wave equation involves the consideration of the radial and angular components of the wavefunction. This is denoted by Eqn. (1.14), where R (r) and A (θ,ϕ) represent the radial and angular wavefunctions, respectively (Yokuş, 2021).

$$\psi_{Cartesian}(x, y, z) = \psi_{radial}(r)\psi_{angular}(\theta, \phi) = R(r)\, A(\theta, \phi) \tag{1.14}$$

1.7. ATOMIC ORBITALS

Within the domain of quantum mechanics, atomic orbitals assume a prominent role as the ethereal platform upon which electrons navigate within an atom. Utilizing these orbitals offers a distinctive and sophisticated method for explaining the conduct and dispersion of electrons, challenging conventional understanding and unveiling the authentic essence of matter on the atomic scale. A comprehensive detail of atomic orbitals is imperative for explaining the intricacies of chemical bonding and the diverse characteristics of matter (Tsaparlis, 1997).

1.7.1. The Quantum Mechanical Description

In classical physics, it is conceivable to envision electrons engaging in orbit around the atomic nucleus, following clearly defined circular trajectories akin to the planetary motion observed in the solar system. Nevertheless, within the area of quantum mechanics, the characteristics exhibited by electrons diverge significantly. The probabilistic framework of quantum mechanics supplants the classical trajectories of electrons.

The Schrödinger Wave Equation postulates that wavefunctions can characterize electrons, which provide information about the likelihood of locating an electron within a specific spatial domain. Atomic orbitals are generated through the solutions to a specific equation, resulting in unique configurations and energy levels that are distinct from one another (Fano, 1957).

1.7.2. Quantum Numbers and Energy Levels

The comprehensive characterization of an atomic orbital necessitates utilizing a collection of quantum numbers. The principal quantum number

(n) determines the primary energy level or shell of the orbital. Higher values of n indicate energy levels that are more elevated and located at greater distances from the nucleus (Vecchi et al., 1976). The determination of the shape of an orbital is governed by the azimuthal quantum number (l), which is intricately linked to the subshells present within a specific energy level. The range of values for the quantum number l is from 0 to (n–1), where n represents the principal quantum number. Each value of l corresponds to a distinct subshell shape, namely s (l=0), p (l=1), d (l=2), and f (l=3). The magnetic quantum number, denoted as ml, is a parameter that determines the spatial orientation of an orbital. Its values span from negative l to positive l. The quantum number n denotes the number of orbitals present in a specific subshell, thereby contributing to the characterization of the spatial arrangement of electron density (Hougen, 1962). The spin quantum number (ms) characterizes the inherent spin of an electron, manifesting as either +1/2 (spin-up) or –1/2 (spin-down). The comprehension of electron pairing and materials' magnetic characteristics heavily relies on this particular characteristic's significance (Mulliken, 1932).

1.7.3. Types of Atomic Orbitals

The four primary categories of atomic orbitals are associated with distinct values of the azimuthal quantum number (l).

1. **s-Orbitals:** These are spherical in shape. These entities possess an orbital angular momentum quantum number of zero (l = 0) and are observed across all energy levels. The likelihood of locating an electron within a s orbital is maximized near the nucleus. It diminishes progressively as one moves away from it in a radial direction.
2. **p-Orbitals:** These exhibit a dumbbell-shaped geometry and are found in groups of three, each corresponding to an angular momentum quantum number (l) of 1. These entities are located in energy levels beyond the initial level and display a dual-lobed structure with a point of zero probability density at the nucleus (Chan et al., 2009).
3. **d-Orbitals:** These exhibit a higher degree of complexity in their shapes, manifesting as five unique orientations that correspond to the quantum number l = 2. Transition metal chemistry involves the participation of these entities, which are found in elevated energy states.

4. **f-Orbitals:** These exhibit a higher level of complexity, as they possess seven distinct orientations that arise from the quantum number l = 3. The behavior of lanthanides and actinides is influenced by these orbitals (Lukens et al., 2013).

1.7.4. Electron Configuration and the Aufbau Principle

The concept of electron configuration pertains to the allocation of electrons within the atomic orbitals of an atom. The principle known as the Aufbau Principle posits that electrons are sequentially distributed into orbitals of progressively lower energy levels prior to occupying orbitals of higher energy.

The electron configurations of atoms adhere to established patterns, resulting in the widely recognized sequence of filling: 1s, 2s, 2p, 3s, 3p, 4s, 3d, and so forth. The comprehension of chemical properties and reactivity of elements, along with the organization of the periodic table, heavily relies on these configurations.

1.7.5. Hybridization and Molecular Geometry

Hybridization of atomic orbitals is a prevalent phenomenon observed in numerous molecules, leading to the formation of novel hybrid orbitals that exhibit distinct geometries and orientations.

The process of hybridization is of utmost importance in the determination of the molecular shape and the angles between chemical bonds, thereby exerting a significant influence on the overall structure and properties of molecules.

Atomic orbitals offer detailed insight into the subatomic realm, wherein the behavior of electrons deviates from classical comprehension. Through the examination of these quantum mechanical dance floors, we are able to acquire significant knowledge regarding the fundamental characteristics of matter and the varied attributes exhibited by elements and compounds.

The comprehension of atomic orbitals serves as a foundation for progress in the fields of chemistry, materials science, and the exploration of the fundamental constituents of the universe. The spectrum of atomic orbitals, ranging from the elementary s orbitals to the intricate f orbitals, harmoniously governs the diverse fabric of matter that envelops our surroundings (et al., 1975).

1.7.6. The Quantum Numbers n, l, and m_l

An atomic orbital is commonly characterized by three discrete quantum numbers that are integral in nature. The principal quantum number, denoted as n, has been previously encountered in the Bohr model of the hydrogen atom. The principal quantum number, denoted as "n," is a positive integer that falls within the range of $1 \leq n \leq \infty$. The permissible values of n are determined through the solution of the radial component of the wave function.

Determining the angular part of the wavefunction introduces two additional quantum numbers, namely l and ml. The quantum number l is commonly called the orbital quantum number and is characterized by a set of permissible values ranging from 0 to (n–1). The determination of the shape of the atomic orbital and the angular momentum of the orbit of the electron is contingent upon the value of l (Leskovec et al., 2019).

Remember

Atoms have the capability to undergo a phenomenon known as "entanglement," wherein the state of one atom has an instantaneous impact on the state of another atom, irrespective of the spatial separation between them.

1.8. THE STRUCTURE OF THE ATOM

According to Dalton's atomic theory, an atom can be defined as the fundamental unit of an element capable of engaging in chemical bonding. Dalton postulated the existence of an atom that possessed two fundamental characteristics: minuteness in size and indivisibility. Nevertheless, a sequence of studies initiated in the 1850s and spanning into the 20th century unequivocally established the presence of internal composition within atoms. This implies that atoms are comprised of smaller entities known as subatomic particles. This study resulted in the identification of three fundamental particles, namely electrons, protons, and neutrons (Thomson, 1904).

1.8.1. The Electron

During the 1890s, a significant number of scientists became involved in the study of radiation, which pertains to the emission and propagation of energy in the form of waves through space. The research findings significantly enhanced our comprehension of atomic structure. A cathode ray tube (CRT), which served as a precursor to the television tube, was employed as a tool to examine this phenomenon. The CRT made of glass undergoes a process of air evacuation, resulting in the removal of a significant portion of its atmospheric content. When the two metal plates are linked to a high-voltage source, the cathode, which is the negatively charged plate, emits an invisible beam. The cathode ray is attracted towards the anode, a positively charged plate, and subsequently traverses through an aperture, persisting in its trajectory towards the opposite extremity of the tube. Upon the incidence of the ray onto the surface that has been coated with a specialized substance, a pronounced fluorescence phenomenon occurs, resulting in the emission of intense luminosity (Laporta, 2018) (Figure 1.3).

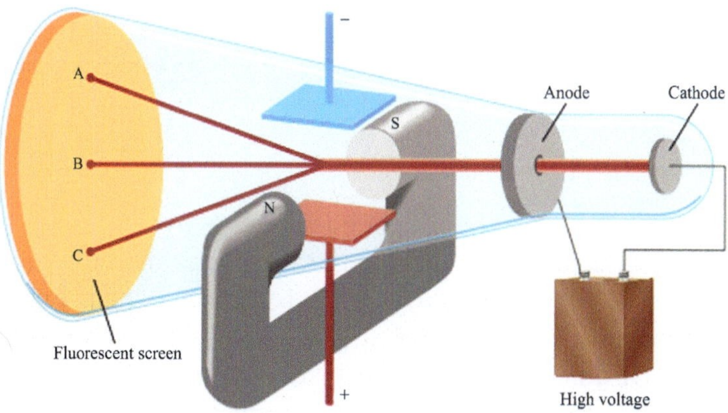

Figure 1.3. *The experimental setup consists of a cathode ray tube, in which an electric field is placed at a right angle to the path of the cathode rays, along with an externally applied magnetic field.*

Source: Stepan, Creative Commons License.

When the magnetic field is energized, and the electric field ceases to exist, the cathode ray comes into contact with point A. When considering solely the influence of the electric field, the ray intersects with point C. When the magnetic and electric fields are simultaneously deactivated or activated in a balanced manner, resulting in the cancellation of their respective influences, the ray will be directed toward point B. The observation can be made, based on the principles of electromagnetic theory, that a moving charged object demonstrates magnetic characteristics and can interact with the magnetic and electric fields it encounters. The attraction of the cathode ray to the plate with positive charges and the repulsion by the plate with negative charges indicates the existence of negatively charged particles within the cathode ray. The entities that exhibit a negative charge are commonly known as electrons. The diagram depicted in Figure 1.4 provides a visual representation of the influence exerted by a bar magnet on the characteristics exhibited by a cathode ray (Ruska, 1987).

Figure 1.4. *The cathode ray, generated within a discharge tube, exhibits a trajectory that extends from the cathode to the anode.*

Source: Stepan, Creative Commons License.

J. J. Thomson, an English physicist, employed a CRT and his understanding of electromagnetic theory to ascertain the ratio between the electric charge and the mass (e/m) of a single electron. The numerical value determined by him is -1.76×10^{8} C/g, wherein the symbol "C" denotes the coulomb, a fundamental unit utilized to quantify electric charge. Subsequently, a sequence of experiments conducted from 1908 to 1917 by R. A. Millikan, an American physicist, yielded the determination that the charge of an electron is -1.6022×10^{-19} C. Based on the provided data, the researcher conducted calculations to determine the mass of an electron (Ackermann, 2007):

$$mass\ of\ an\ electron = \frac{charge}{\frac{e}{m}} = \times \frac{-1.6022 \times 10^{-19} C}{-1.76 \times 10^{8} C/g} = 9.10 \times 10 - 28\ \mathrm{g}$$

1.8.2. Radioactivity

In 1895, Wilhelm Röntgen, a physicist hailing from Germany, observed the effects of cathode rays on glass and metals, which resulted in the emission of highly peculiar rays. This radiation, characterized by its significant energy levels, exhibited the ability to permeate through matter, resulting in the darkening of photographic plates that were shielded, as well as inducing fluorescence in a diverse range of substances.

Due to their inability to be deflected by a magnet, it can be inferred that these rays do not possess charged particles akin to cathode rays. X-rays were named by Röntgen. Shortly following Röntgen's groundbreaking exploration, Antoine Becquerel, a distinguished professor of physics-based in Paris, investigated the fluorescent characteristics exhibited by various substances (Boisdron & Brock, 1970). Serendipitously, he fortuitously discovered that subjecting heavily encased photographic plates to a specific uranium compound resulted in their darkening, even in the absence of cathode ray stimulation. Similar to X-rays, the rays emitted by the uranium compound possessed significant energy and were impervious to magnetic deflection. However, in contrast to X-rays, these rays were generated in a spontaneous manner. Marie Curie, a student of Becquerel, proposed the term "radioactivity" to denote the phenomenon of natural release of particles and/or radiation. Hence, any element that exhibits the phenomenon of emitting radiation without external stimulation is referred to as radioactive. Subsequent examination has unveiled that the decay process of radioactive materials, such as uranium, produces three distinct categories of rays. Two out of the three types are redirected when encountering metal plates with opposite charges. Alpha (α) rays are composed of positively charged particles known as alpha particles, and as a result, they experience deflection when encountering a positively charged plate. Beta (β) rays, also known as beta particles, refer to electrons that exhibit deflection when subjected to a negatively charged plate. The third category of radioactive radiation encompasses gamma (g) rays, which are characterized by their high energy levels (Prather, 2005) (Figure 1.5).

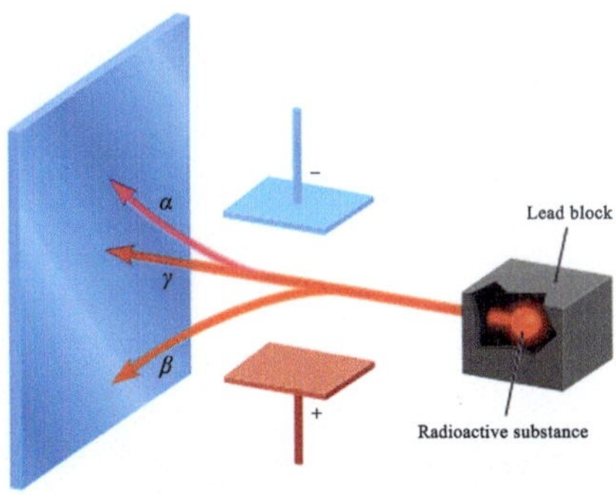

Figure 1.5. *Three distinct categories of rays are emitted by radioactive elements.*

Source: Stepan, Creative Commons License.

1.8.3. Proton and Nucleus

During the initial years of the 20th century, it became evident that atoms possess two distinct characteristics: the presence of electrons and their overall electrical neutrality. In order to preserve electrical neutrality, an atom must possess an equivalent quantity of both positive and negative charges. Based on the provided information, Thomson postulated the concept of an atom as a homogeneous, positively charged entity encompassing embedded electrons (*see* Figure 1.6). Thomson's widely recognized "plum pudding" model was the prevailing theory for a considerable period of time. In the year 1910, Ernest Rutherford, a physicist hailing from New Zealand, made the decision to employ particles as a means to investigate the composition of atoms. Before this, Rutherford pursued studies under Thomson's tutelage at the esteemed Cambridge University (Dusling & Venugopalan, 2013).

The researchers made an observation that a significant proportion of particles were able to pass through the foil without any deviation or with only a minimal deviation. Additionally, it was observed that sporadically, an individual particle underwent scattering or deflection at a significant angle. In certain cases, a particle exhibited the phenomenon of rebounding in the opposite direction of its initial trajectory.

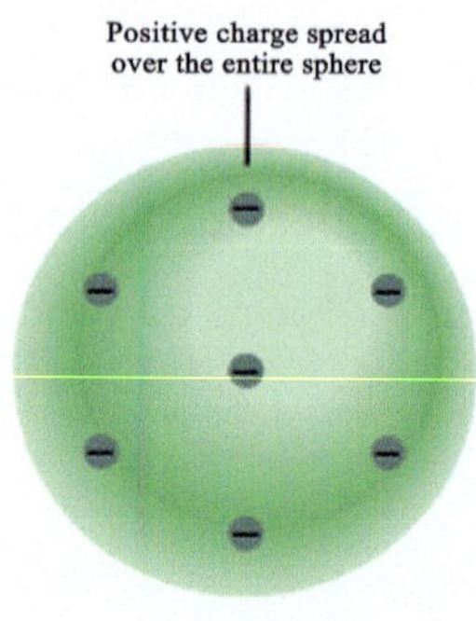

Figure 1.6. *Thomson's atomic model, commonly known as the "plum-pudding" model.*

Source: David, Creative Commons License.

The discovery was highly unexpected, as Thomson's model proposed that the positive charge of the atom was widely dispersed, resulting in minimal deflection of the positive alpha particles (Badhwar et al., 1977). According to Rutherford's initial response upon being informed of this discovery, he expressed astonishment by comparing it to the implausible scenario of firing a 15-inch shell at an area of tissue paper, only to have it rebound and strike the shooter. In order to discuss the outcomes of the a-scattering experiment, Rutherford formulated a novel conceptual framework pertaining to atomic structure, positing that a significant portion of the atom is comprised of vacant space (Bzdak et al., 2013) (Figure 1.7).

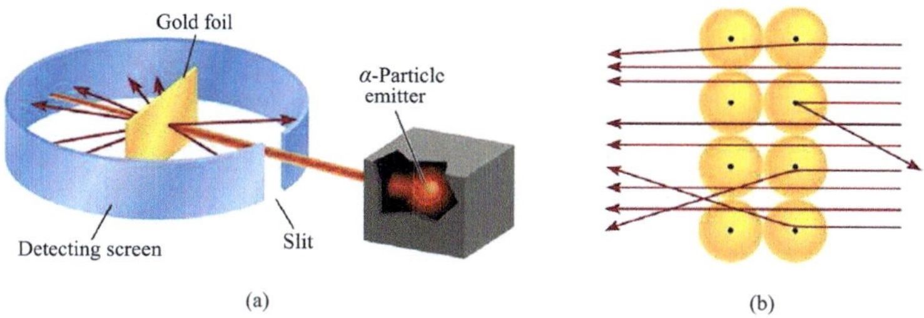

Figure 1.7. *Rutherford's gold foil experiment – (a) a magnified observation of particles traversing through; and (b) experiencing deflection due to interactions with atomic nuclei.*

Source: David W., Creative Commons License.

According to Rutherford's proposal, the positive charges of an atom are localized within the nucleus, which is a highly compact central region of the atom. In the context of the scattering experiment, it was observed that whenever a particle approached the nucleus, it encountered a significant repulsive force, resulting in a substantial deflection. In addition, it should be noted that a particle moving in a direct trajectory toward a nucleus would encounter a significant repulsive force, potentially leading to a complete reversal of its direction. The particles within the nucleus that possess a positive charge are referred to as protons. In distinct experimental trials, it was observed that the electric charge of each proton possesses an equivalent magnitude to that of an electron (Yan & Ollitrault, 2014).

Additionally, it was determined that the mass of a proton is approximately 1.67262×10^{-24} g, which is roughly 1840 times greater than the mass of an electron, which carries an opposite charge. During this investigation phase, scientists held the perception of the atom in the following manner. The majority of an atom's mass is attributed to its nucleus, even though the nucleus occupies a mere fraction of the atom's volume, approximately $1/10^{13}$. Atomic dimensions are typically quantified using the picometer (pm), the standard unit of measurement in the International System of Units (Rehm et al., 1998).

$$1pm = 1 \times 10^{-12}m$$

1.8.4. The Neutron

Rutherford's atomic structure model failed to address a significant unresolved issue. It was widely recognized that hydrogen, the most elementary atom, consists solely of one proton, while the helium atom is composed of two protons. Consequently, the proportion between the mass of a helium atom and that of a hydrogen atom is expected to be 2:1. Due to the significant disparity in mass between electrons and protons, the impact of electrons can be disregarded. In actuality, the ratio stands at 4:1. Rutherford and his contemporaries posited the existence of an additional subatomic particle within the atomic nucleus.

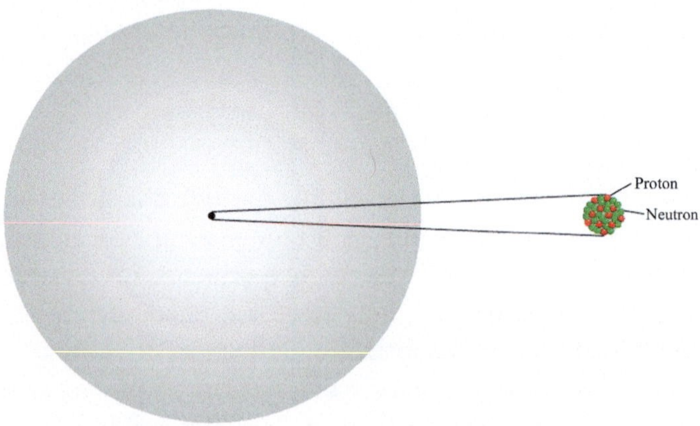

Figure 1.8. *Densely concentrated neutrons and protons in an atom are within a tiny nucleus.*

Source: David W., Creative Commons License.

This proposition was subsequently substantiated by James Chadwick, an English physicist, in the year 1932. When Chadwick subjected a thin sheet of beryllium to bombardment by alpha particles, a form of high-energy radiation resembling gamma rays was emitted by the metal. Subsequent experiments provided evidence indicating that the observed rays were comprised of electrically uncharged particles with a mass slightly surpassing that of protons. The discovered were designated as neutrons by Chadwick. The explanation of the mass ratio conundrum is now attainable. The helium nucleus consists of two protons and two neutrons, while the hydrogen nucleus contains only one proton and no neutrons. The diagram presented in Figure 1.8 illustrates the spatial arrangement of elementary particles, namely protons, neutrons, and electrons, within an atom. In the domain of subatomic particles, it is worth noting that while there exist additional entities, the electron, proton, and neutron hold paramount significance as the three fundamental constituents of the atom, particularly within the domain of chemistry (Howell et al., 1998).

1.9. THE PERIODIC TABLE

A significant proportion of the elements currently recognized were identified during the period spanning from 1800 to 1900. During this particular era, chemists observed that numerous elements exhibit significant resemblances to one another. The idea for the periodic table came from the fact that physical and biological changes tend to happen in cycles. Because of this, a chart was made to group together elements with similar physical and chemical traits. The modern periodic table, as depicted in Figure 1.9, exhibits the arrangement of elements based on their atomic number, denoted above the respective element symbol. This arrangement organizes the elements into periods, which are horizontal rows, and groups, which are vertical columns. The grouping is determined by the similarities observed in the chemical properties of the elements. The synthesis of elements 112, 114, 116, and 118 has been achieved in recent times, albeit without assigned names at present (Lengler & Eppler, 2007).

The elements can be classified into three distinct categories, namely metals, nonmetals, and metalloids. Metals are known for their high thermal and electrical conductivity, while nonmetals generally exhibit low thermal and electrical conductivity. Metalloid exhibits characteristic that lies between those of metals and nonmetals. According to the data presented in Figure 1.9, it can be observed that the predominant category among the known elements is metals. At the same time, nonmetals comprise a smaller portion with a count of 17 elements.

Additionally, metalloids, a distinct subset, are represented by eight elements. Across any given period, the elements exhibit a gradual transition in their physical and chemical properties from a metallic nature to a nonmetallic nature when moving from left to right. The periodic table serves as a valuable instrument that systematically corresponds to the properties of elements, enabling us to make informed predictions regarding their chemical behavior.

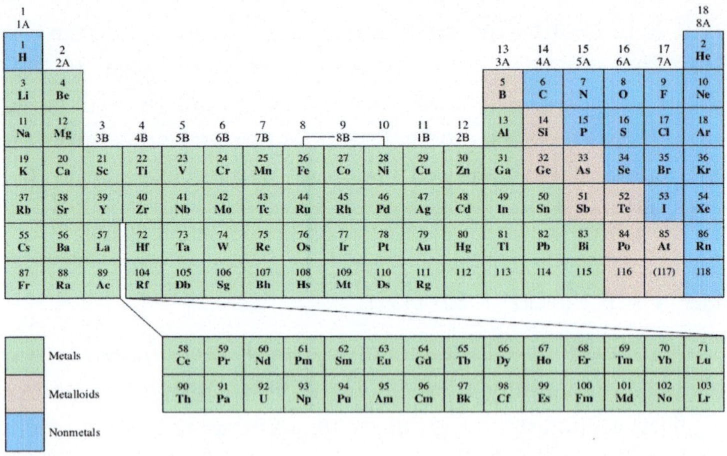

Figure 1.9. *The contemporary iteration of the periodic table.*

Source: Raymond, Creative Commons License.

Elements are commonly denoted collectively based on their respective group numbers in the periodic table, such as Group 1A, Group 2A, and so forth. However, in the interest of convenience, certain groups of elements have been assigned specific names. The elements found in Group 1A, namely Li, Na, K, Rb, Cs, and Fr, are commonly referred to as alkali metals. Similarly, the elements present in Group 2A, namely Be, Mg, Ca, Sr, Ba, and Ra, are commonly known as alkaline earth metals. The elements comprising Group 7A, namely fluorine (F), chlorine (Cl), bromine (Br), iodine (I), and astatine (At), are commonly referred to as halogens. Conversely, the elements found in Group 8A, including helium (He), neon (Ne), argon (Ar), krypton (Kr), xenon (Xe), and radon (Rn), are commonly known as noble gases or rare gases (Jena, 2013).

According to the established periodic table (Figure 1.9), it is observed that cerium (Ce) is positioned subsequent to lanthanum (La), while thorium (Th) is situated immediately following actinium (Ac). The International Union of Pure and Applied Chemistry (IUPAC) has made a recommendation for the use of the 1–18 group designation, although its adoption has not yet been widely implemented. In this passage, the conventional U.S. notation for group numbers (1A–8A and 1B–8B) is employed. The elements with atomic numbers 112–116 and 118 have not yet been assigned specific names. The synthesis of Element 117 has not been achieved thus far.

Did You Know?

The quantity of atoms contained within an individual grain of sand is approximately equivalent to the quantity of grains of sand present on the beaches of planet Earth.

1.10. NUCLEAR CHEMISTRY

Nuclear chemistry is an intriguing scientific field investigating the characteristics, alterations, and practical uses of atomic nuclei. The foundation of nuclear chemistry is rooted in the comprehension of nuclear reactions, radioactivity, and the fundamental forces that dictate the actions of subatomic particles. The nuclear chemistry field has significantly influenced multiple domains, encompassing energy generation, medical applications, and the advancement of our comprehension of the cosmos. This chapter aims to explore the fundamental elements of nuclear chemistry, encompassing the historical breakthrough of radioactivity and its wide-ranging practical implementations (Tekin & Nakiboglu, 2006).

1.10.1. Discovery of Radioactivity

The exploration of nuclear chemistry commenced in the late 19th century as scientists delved into the properties of matter and its fundamental components, leading to the discovery of radioactivity. The accidental discovery of radioactivity occurred in 1896 when French physicist Henri Becquerel was conducting research on the impacts of X-rays. The researcher observed that specific uranium salts emitted radiation capable of penetrating black paper and sensitizing photographic plates, even without light exposure. The fortuitous nature of this discovery served as the basis for subsequent inquiries into the enigmatic phenomenon of radioactivity (Kipnis, 2000).

1.10.2. Types of Radioactive Decay

The phenomenon of radioactive decay pertains to the inherent and autonomous alteration of an unsteady atomic nucleus into a state of greater stability through the emission of radiation. There exist three fundamental categories of radioactive decay:

1. **Alpha Decay:** It is a nuclear process wherein an atomic nucleus releases an alpha particle, which is composed of two protons and two neutrons and is equivalent to a helium nucleus. The resultant daughter nucleus exhibits a decrease in atomic number by two units compared to the parent nucleus, along with a reduction in mass number by four units. The process of alpha decay can be represented in the following manner (Sandulescu et al., 1980):

$$\text{Parent nucleus } (Z, A) \rightarrow \text{Daughter nucleus } (Z-2, A-4) + \alpha \text{ particle } (2, 4)$$

2. **Beta Decay:** It occurs in two forms: beta-minus decay (β-) and beta-plus decay (β+).

a. **Beta-Minus Decay:** It is a nuclear process wherein a neutron within the nucleus converts into a proton, accompanied by the emission of an electron and an antineutrino. The atomic number of the daughter nucleus is incremented by one compared to the parent nucleus, while the mass number remains unchanged. The process of beta-minus decay can be mathematically represented as:

$$Parent\ nucleus\ (Z, A) \rightarrow Daughter\ nucleus\ (Z+1, A) + \beta^{-}\ particle\ (-1, 0) + \text{antineutrino}$$

b. **Beta-Plus Decay:** It is a nuclear process wherein a proton within the nucleus undergoes conversion into a neutron. This conversion is accompanied by the emission of a positron, which is a positively charged electron, as well as a neutrino. The daughter nucleus exhibits an atomic number that is decreased by one compared to the parent nucleus, while the mass number remains unchanged. The process of beta-plus decay can be mathematically represented as:

$$Parent\ nucleus\ (Z, A) \rightarrow Daughter\ nucleus\ (Z-1, A) + \beta^{+}\ particle\ (+1, 0) + neutrino$$

3. **Gamma Decay:** Following the process of alpha or beta decay, it is possible for the resultant daughter nucleus to remain in an energetically excited state. In order to achieve a state of greater energy stability, an entity discharges surplus energy in the form of electromagnetic radiation known as gamma rays. Gamma rays possess neither mass nor charge (Streicher et al., 2010).

1.10.3. Nuclear Reactions

Nuclear reactions encompass the reconfiguration of atomic nuclei via nuclear fission and fusion mechanisms.

1. **Nuclear Fission:** It refers to the phenomenon in which a nucleus with significant mass, such as uranium-235 or plutonium-239, undergoes division into two or more nuclei of lesser mass, resulting in the liberation of a substantial quantity of energy. This process serves as the fundamental principle for the generation of nuclear power and the development of nuclear weapons. The fission reaction may be represented as (Fraser & Milton, 1966):

$$Nucleus\ (Z, A) + Neutron \rightarrow Fission\ Fragments + Neutrons + Energy$$

2. **Nuclear Fusion:** It is the phenomenon in which two light atomic nuclei are merged, resulting in the formation of a nucleus with greater mass. The process above is responsible for generating energy in celestial bodies such as the sun and other stars. Moreover, it exhibits considerable potential in terms of providing a sustainable and plentiful energy source for our planet. Nuclear fusion can be observed in the fusion reaction involving hydrogen isotopes, specifically deuterium and tritium, which produce helium and a neutron (Rieth et al., 2013).

$$Deuterium\ (1, 2) + Tritium\ (1, 3) \rightarrow Helium\ (2, 4) + Neutron$$

1.10.4. Radioactive Decay Kinetics

The phenomenon of radioactive decay adheres to the principles of first-order kinetics, wherein the rate of decay is directly proportional to the quantity of radioactive nuclei in existence. The process of radioactive decay can be mathematically characterized by the utilization of the radioactive decay law (McNair et al., 1969):

$$N(t) = N^0 x\, e^{-\lambda t} \qquad (1.15)$$

where; the variable N(t) represents the quantity of radioactive nuclei at a given time 't;' N_0 denotes the initial quantity of radioactive nuclei, while λ represents the decay constant, which is a unique property of the radioactive isotope. e is the base of the natural logarithm, which is equal to 2.718. The half-life (t½) of a radioactive isotope refers to the temporal duration required for half of the initial sample to undergo decay. The calculation can be determined as follows:

$$t½ = ln(2) / \lambda \quad (1.16)$$

The understanding of half-life plays a pivotal role in the field of radiometric dating, enabling scientists to ascertain the chronological age of historic objects and geological samples (Metzger et al., 2010).

1.10.5. Applications of Nuclear Chemistry

Nuclear chemistry has assumed a crucial role in a multitude of practical uses that have yielded substantial benefits to society (Marcone et al., 2013).

1. **Nuclear Power Generation:** Nuclear power plants utilize the energy released through nuclear fission reactions in order to generate electrical power. The process exhibits high efficiency, and nuclear power offers a substantial, environmentally friendly energy source on a large scale (Yim & Caron, 2006).
2. **Radioisotopes in Medicine:** The utilization of radioactive isotopes is widespread in the field of medical diagnostics and treatment (Luo et al., 2016). Positron emission tomography (PET) diagnostic techniques utilize radioisotopes to facilitate the visualization of metabolic processes within the human body. Moreover, radioisotopes find application in the field of radiation therapy for the purpose of cancer treatment (Richards, 1960).
3. **Radiocarbon Dating:** It is a scientific method that utilizes the decay of carbon-14 to ascertain the chronological age of organic substances (Zazzo & Saliège, 2011).
4. **Industrial Applications:** Nuclear techniques find application in various industries for the purpose of process monitoring, thickness measuring, and material analysis (Madhavan et al., 2017).
5. **Nuclear Imaging and Research:** Nuclear imaging techniques, such as gamma-ray spectroscopy and neutron activation analysis, are employed in scientific research and analysis (Chauveau et al., 2008).

1.10.6. Challenges and Safety

Despite the myriad advantages, nuclear chemistry also presents various challenges and safety considerations. Ensuring the appropriate management, containment, and elimination of radioactive substances is of utmost importance to mitigate the risk of accidents and potential adverse effects on human health and the environment. The effective handling and disposal of nuclear waste continues to be a multifaceted matter necessitating continuous investigation and focus (Schenk et al., 2018).

The field of nuclear chemistry has played a pivotal role in shaping the contemporary world and remains highly promising in addressing global challenges such as energy requirements and advancements in medical science (Karmaus et al., 2018).

The comprehension of nuclear reactions and radioactivity has facilitated the exploration of diverse applications that confer advantages to the human population. As the investigation into nuclear chemistry progresses, it becomes imperative to place emphasis on the prioritization of safety and responsible handling of nuclear materials in order to guarantee a future that is both sustainable and secure (Wahsner et al., 2018).

Did You Know?

The term "atom" originates from the ancient Greek term "atomos," which denotes the concept of indivisibility or the inability to be cut.

1.11. SUMMARY

The chapter offers a comprehensive exploration of the core aspects of atomic structure and behavior. The introductory section provides an overview of the elementary constituents of an atom, namely protons, neutrons, and electrons. These particles constitute the fundamental building blocks of all matter, and their interplay governs the distinctive characteristics exhibited by various elements. The fundamental understanding of the arrangement of elements in the periodic table heavily relies on the concept of atomic number, which signifies the number of protons within an atom, and mass number, which denotes the combined count of protons and neutrons. Furthermore, the presence of isotopes, which are atoms of the same element but with varying numbers of neutrons, provides insight into the variability of atoms at the atomic level.

The chapter extensively examines Bohr's Theory of the Atomic Spectrum of Hydrogen, which is considered a fundamental aspect of atomic theory. The proposed model significantly transformed our comprehension of atomic structure by positing the existence of discrete energy levels in which electrons orbit the nucleus. Moreover, it discussed that electrons have the capacity to absorb or emit energy in the form of photons, thereby facilitating transitions between these energy levels. The presented model effectively discusses the spectral lines seen in the hydrogen spectrum, providing a profound understanding of the discrete nature of energy levels.

In conclusion, the chapter provides a comprehensive discussion about the atomic structure and the dynamics exhibited by subatomic entities. It plays a fundamental role in comprehending the realm of quantum mechanics and nuclear chemistry, thereby facilitating progress in diverse scientific fields and technological innovations.

REFERENCES

1. Abdelghafar, F., Xu, X., & Shao, Z., (2022). Designing single-atom catalysts toward improved alkaline hydrogen evolution reaction. *Materials Reports: Energy*, *10*(3), 100144.
2. Ackermann, H. W., (2007). 5500 Phages examined in the electron microscope. *Archives of Virology*, *152*, 227–243.
3. Alsaif, N. A., Radiman, S., & Ahmed, S. M. S., (2017). Calculations of alpha cluster preformation probability for the three even–even superheavy isotopes with atomic number Z= 108, 110, and 112. *International Journal of Modern Physics E*, *26*(03), 1750008.
4. Badhwar, G. D., Stephens, S. A., & Golden, R. L., (1977). Analytic representation of the proton-proton and proton-nucleus cross-sections and its application to the sea-level spectrum and charge ratio of muons. *Physical Review D*, *15*(3), 820.
5. Barry, N. P., & Sadler, P. J., (2013). Exploration of the medical periodic table: Towards new targets. *Chemical Communications*, *49*(45), 5106–5131.
6. Benczik, S., Chang, L. N., Minic, D., & Takeuchi, T., (2005). Hydrogen-atom spectrum under a minimal-length hypothesis. *Physical Review A*, *72*(1), 012104.
7. Block, M., Ackermann, D., Blaum, K., Droese, C., Dworschak, M., Eliseev, S., & Weber, C., (2010). Direct mass measurements above uranium bridge the gap to the island of stability. *Nature*, *463*(7282), 785–788.
8. Bohr, N., (1913). I. On the constitution of atoms and molecules. *The London, Edinburgh, and Dublin Philosophical Magazine and Journal of Science*, *26*(151), 1–25.
9. Boisdron, Y., & Brock, J. R., (1970). On the stochastic nature of the acquisition of electrical charge and radioactivity by aerosol particles. *Atmospheric Environment (1967)*, *4*(1), 35–50.
10. Bzdak, A., Schenke, B., Tribedy, P., & Venugopalan, R., (2013). Initial-state geometry and the role of hydrodynamics in proton-proton, proton-nucleus, and deuteron-nucleus collisions. *Physical Review C*, *87*(6), 064906.
11. Chan, J. A., Lany, S., & Zunger, A., (2009). Electronic correlation in anion p orbitals impedes ferromagnetism due to cation vacancies in Zn chalcogenides. *Physical Review Letters*, *103*(1), 016404.

12. Charalampous, C., (2021). The confined atom: James Clerk Maxwell on the fundamental particles and the limits of scientific knowledge. *Perspectives on Science*, *29*(2), 189–214.
13. Chauveau, F., Boutin, H., Van Camp, N., Dollé, F., & Tavitian, B., (2008). Nuclear imaging of neuroinflammation: A comprehensive review of [11 C] PK11195 challengers. *European Journal of Nuclear Medicine and Molecular Imaging*, *35*, 2304–2319.
14. Coessens, V. M., & Matyjaszewski, K., (2010). Fundamentals of atom transfer radical polymerization. *Journal of Chemical Education*, *87*(9), 916–919.
15. Cui, T., Li, L., Ye, C., Li, X., Liu, C., Zhu, S., & Wang, D., (2022). Heterogeneous single atom environmental catalysis: Fundamentals, applications, and opportunities. *Advanced Functional Materials*, *32*(9), 2108381.
16. Dini, V., & Hammer, D., (2017). Case study of a successful learner's epistemological framings of quantum mechanics. *Physical Review Physics Education Research*, *13*(1), 010124.
17. Domenicano, A., Vaciago, A., & Coulson, C. A., (1975). Molecular geometry of substituted benzene derivatives. I. On the nature of the ring deformations induced by substitution. *Acta Crystallographica Section B: Structural Crystallography and Crystal Chemistry*, *31*(1), 221–234.
18. Dorai, K., Chaturvedi, S., & Mukunda, N., (2018). The development of quantum mechanics. *Resonance: Journal of Science Education*, *23*(10), 5841–5845.
19. Dusling, K., & Venugopalan, R., (2013). Comparison of the color glass condensate to dihadron correlations in proton-proton and proton-nucleus collisions. *Physical Review D*, *87*(9), 094034.
20. Fano, U., (1957). Description of states in quantum mechanics by density matrix and operator techniques. *Reviews of Modern Physics*, *29*(1), 74.
21. Firestone, R. B., & Shirley, V. S., (1998). *Table of Isotopes, 2 Volume Set* (Vol. 2, pp. 3168).
22. Fraser, J. S., & Milton, J. C. D., (1966). Nuclear fission. *Annual Review of Nuclear Science*, *16*(1), 379–444.
23. Gao, J., Luo, X., Fang, F., & Sun, J., (2021). Fundamentals of atomic and close-to-atomic scale manufacturing: a review. *International Journal of Extreme Manufacturing*, *4*(1), 012001.

24. Gearhart, C. A., (2010). "Astonishing successes" and "bitter disappointment": The specific heat of hydrogen in quantum theory. *Archive for History of Exact Sciences*, *64*, 113–202.
25. Glauber, R. J., (1963). The quantum theory of optical coherence. *Physical Review*, *130*(6), 2529.
26. Hougen, J. T., (1962). Classification of rotational energy levels for symmetric - top molecules. *The Journal of Chemical Physics*, *37*(7), 1433–1441.
27. Howell, C. R., Chen, Q., Carman, T. S., Hussein, A., Gibbs, W. R., Gibson, B. F., & Whitton, M., (1998). Toward a resolution of the neutron-neutron scattering-length issue. *Physics Letters B*, *444*(3–4), 252–259.
28. Jena, P., (2013). Beyond the periodic table of elements: The role of superatoms. *The Journal of Physical Chemistry Letters*, *4*(9), 1432–1442.
29. Karmaus, A. L., Osborn, R., & Krishan, M., (2018). Scientific advances and challenges in safety evaluation of food packaging materials: Workshop proceedings. *Regulatory Toxicology and Pharmacology*, *98*, 80–87.
30. Khandelwal, A., Maarisetty, D., & Baral, S. S., (2022). Fundamentals and application of single atom photocatalyst in sustainable energy and environmental applications. *Renewable and Sustainable Energy Reviews*, *167*, 112693.
31. King, J. F., & Fall, P. H., (1940). Radioactivity and the periodic table: An introduction to the study of atomic structure and isotopes in general chemistry. *Journal of Chemical Education*, *17*(10), 481.
32. Kipnis, N. S., (2000). The window of opportunity: Logic and chance in Becquerel's discovery of radioactivity. *Physics in Perspective*, *2*, 63–99.
33. Kochen, S., & Specker, E. P., (1990). The problem of hidden variables in quantum mechanics. *Ernst Specker Selecta*, *4*, 235–263.
34. Kratz, J. V., Loveland, W., & Moody, K. J., (2015). Syntheses of transuranium isotopes with atomic numbers Z≤ 103 in multi-nucleon transfer reactions. *Nuclear Physics A*, *944*, 117–157.
35. Kurudirek, M., (2017). Effective atomic number of soft tissues, water, and air for interaction of various hadrons, leptons, and isotopes of hydrogen. *International Journal of Radiation Biology*, *93*(12), 1299–1305.

36. Laporta, S., (2018). High-precision calculation of the 4-loop contribution to the electron g-2 in QED1. In: *Journal of Physics: Conference Series* (Vol. 1085, pp. 52008–52009). IOP Publishing.
37. Lengler, R., & Eppler, M. J., (2007). Towards a periodic table of visualization methods for management. In: *IASTED Proceedings of the Conference on Graphics and Visualization in Engineering (GVE 2007), Clearwater, Florida, USA* (Vol. 7, pp. 31–32).
38. Leskovec, L., Meinel, S., Pflaumer, M., & Wagner, M., (2019). Lattice QCD investigation of a doubly-bottom *ud* tetraquark with quantum numbers $I(J^P) = 0(1^+)$. *Physical Review D, 100*(1), 014503.
39. Li, Y., Zhou, Y., Shang, C., Yousaf, M., Guo, Z., & Guo, S., (2022). Emerging metal single-atom materials: From fundamentals to energy applications. *Accounts of Materials Research, 3*(11), 1160–1172.
40. Lodders, K., (2021). Relative atomic solar system abundances, mass fractions, and atomic masses of the elements and their isotopes, composition of the solar photosphere, and compositions of the major chondritic meteorite groups. *Space Science Reviews, 217*, 1–33.
41. Loncar, S. J., (2021). Science and religion: AN origins story: With James C. Ungureanu, "Introduction to the Symposium on Science, Religion, and the Rise of Biblical Criticism"; Paul CH Lim, "Atheism, Atoms, and Activity of God: Science and Religion in Early Boyle Lectures, 1692–1720"; Diego Lucci, "The Biblical Roots of Locke's Theory of Personal Identity"; Jon W. Thompson, "The Naturalization of Scriptural Reason in 17th Century Epistemology"; James C. Ungureanu, 'From Divine Oracles to the Higher Criticism': Andrew D. *Zygon®, 56*(1), 275–296.
42. Lukens, W. W., Edelstein, N. M., Magnani, N., Hayton, T. W., Fortier, S., & Seaman, L. A., (2013). Quantifying the σ and π interactions between U (V) f orbitals and halide, alkyl, alkoxide, amide, and ketimide ligands. *Journal of the American Chemical Society, 135*(29), 10742–10754.
43. Luo, W., Bobeica, M., Gheorghe, I., Filipescu, D. M., Niculae, D., & Balabanski, D. L., (2016). Estimates for production of radioisotopes of medical interest at extreme light infrastructure – nuclear physics facility. *Applied Physics B, 122*, 1–11.
44. Madhavan, A., Sindhu, R., Binod, P., Sukumaran, R. K., & Pandey, A., (2017). Strategies for design of improved biocatalysts for industrial applications. *Bioresource Technology, 245*, 1304–1313.

45. Marcone, M. F., Wang, S., Albabish, W., Nie, S., Somnarain, D., & Hill, A., (2013). Diverse food-based applications of nuclear magnetic resonance (NMR) technology. *Food Research International, 51*(2), 729–747.

46. Matyjaszewski, K., & Xia, J., (2002). Fundamentals of atom transfer radical polymerization. *Handbook of Radical Polymerization*, 523–628.

47. McNair, A., Bannister, F. J., Keith, R. L. G., & Wilson, H. W., (1969). A measurement of the energy released as kinetic energy of β-particles emitted in the radioactive decay of the fission products of 235U. *Journal of Nuclear Energy, 23*(2), 73–86.

48. Metzger, B. D., Martínez-Pinedo, G., Darbha, S., Quataert, E., Arcones, A., Kasen, D., & Zinner, N. T., (2010). Electromagnetic counterparts of compact object mergers powered by the radioactive decay of r-process nuclei. *Monthly Notices of the Royal Astronomical Society, 406*(4), 2650–2662.

49. Mills, R. L., (2000). The hydrogen atom revisited. *International Journal of Hydrogen Energy, 25*(12), 1171–1183.

50. Moumni, M., BenSlama, A., & Zaim, S., (2010). Spectrum of hydrogen atom in space-time non-commutativity. *arXiv Preprint arXiv, 40*(3), 1003–1005.

51. Mulliken, R. S., (1932). The interpretation of band spectra part III. Electron quantum numbers and states of molecules and their atoms. *Reviews of Modern Physics, 4*(1), 1.

52. Nguyen, T. N., Salehi, M., Le, Q. V., Seifitokaldani, A., & Dinh, C. T., (2020). Fundamentals of electrochemical CO_2 reduction on single-metal-atom catalysts. *ACS Catalysis, 10*(17), 10068–10095.

53. Oganessian, Y. T., Abdullin, F. S., Bailey, P. D., Benker, D. E., Bennett, M. E., Dmitriev, S. N., & Wilk, P. A., (2010). Synthesis of a new element with atomic number Z= 117. *Physical Review Letters, 104*(14), 142502.

54. Peterson, B. J., & Fry, B., (1987). Stable isotopes in ecosystem studies. *Annual Review of Ecology and Systematics, 18*(1), 293–320.

55. Prather, E., (2005). Students' beliefs about the role of atoms in radioactive decay and half-life. *Journal of Geoscience Education, 53*(4), 345–354.

56. Rehm, K. E., Esbensen, H., Jiang, C. L., Back, B. B., Borasi, F., Harss, B., & Wuosmaa, A. H., (1998). Fusion cross sections for the proton drip line nucleus F 17 at energies below the coulomb barrier. *Physical Review Letters*, *81*(16), 3341.

57. Richards, P., (1960). A survey of the production at Brookhaven National Laboratory of radioisotopes for medical research. In: *Fifth Nuclear Congress, Rome, Italy* (Vol. 2, pp. 41–52).

58. Rieth, M., Dudarev, S. L., De Vicente, S. G., Aktaa, J., Ahlgren, T., Antusch, S., & Zivelonghi, A., (2013). Recent progress in research on tungsten materials for nuclear fusion applications in Europe. *Journal of Nuclear Materials*, *432*(1–3), 482–500.

59. Rodríguez, M. A., & Niaz, M., (2002). How in spite of the rhetoric, history of chemistry has been ignored in presenting atomic structure in textbooks. *Science & Education*, *11*, 423–441.

60. Roy, S. S., (2016). Simplification of master power expression and effective power detection of QCA device (wave nature tunneling of electron in QCA device). In: *2016 IEEE Students' Technology Symposium (TechSym)* (Vol. 2, pp. 272–277). IEEE.

61. Ruska, E., (1987). The development of the electron microscope and of electron microscopy. *Reviews of Modern Physics*, *59*(3), 627.

62. Salvat, F., Fernández-Varea, J. M., & Williamson, Jr. W., (1995). Accurate numerical solution of the radial Schrödinger and Dirac wave equations. *Computer Physics Communications*, *90*(1), 151–168.

63. Sandulescu, A., Poenaru, D. N., & Greiner, W., (1980). New type of decay of heavy nuclei intermediates between fission and. cap alpha. decay. *Sov. J. Particles Nucl. (Engl. Transl.); (United States)*, *11*(6), 987–990.

64. Schenk, L., Taher, I. A., & Öberg, M., (2018). Identifying the scope of safety issues and challenges to safety management in Swedish middle school and high school chemistry education. *Journal of Chemical Education*, *95*(7), 1132–1139.

65. Stetsko, M. M., & Tkachuk, V. M., (2006). Perturbation hydrogen-atom spectrum in deformed space with minimal length. *Physical Review A*, *74*(1), 012101.

66. Streicher, B., Heßberger, F. P., Antalic, S., Hofmann, S., Ackermann, D., Heinz, S., & Venhart, M., (2010). Alpha-gamma decay studies of

261Sg and 257Rf. *The European Physical Journal A*, *45*(3), 275–286.

67. Tekin, B. B., & Nakiboglu, C., (2006). Identifying students' misconceptions about nuclear chemistry. A study of Turkish high school students. *Journal of Chemical Education*, *83*(11), 1712.
68. Thomson, J. J., (1904). XXIV. On the structure of the atom: An investigation of the stability and periods of oscillation of a number of corpuscles arranged at equal intervals around the circumference of a circle; with application of the results to the theory of atomic structure. *The London, Edinburgh, and Dublin Philosophical Magazine and Journal of Science*, *7*(39), 237–265.
69. Tsaparlis, G., (1997). Atomic orbitals, molecular orbitals and related concepts: Conceptual difficulties among chemistry students. *Research in Science Education*, *27*, 271–287.
70. Vecchi, M. P., Pereira, J. R., & Dresselhaus, M. S., (1976). Anomalies in the magnetoreflection spectrum of bismuth in the low-quantum-number limit. *Physical Review B*, *14*(2), 298.
71. Viana, H. E. B., & Porto, P. A., (2010). The development of Dalton's atomic theory as a case study in the history of science: Reflections for educators in chemistry. *Science & Education*, *19*, 75–90.
72. Wahsner, J., Gale, E. M., Rodríguez-Rodríguez, A., & Caravan, P., (2018). Chemistry of MRI contrast agents: Current challenges and new frontiers. *Chemical Reviews*, *119*(2), 957–1057.
73. Wang, X., Zhang, Y., Wu, J., Zhang, Z., Liao, Q., Kang, Z., & Zhang, Y., (2021). Single-atom engineering to ignite 2D transition metal dichalcogenide based catalysis: Fundamentals, progress, and beyond. *Chemical Reviews*, *122*(1), 1273–1348.
74. Weinert, F., (2000). The construction of atom models: Eliminative inductivism and its relation to falsificationism. *Foundations of Science*, *5*, 491–531.
75. Woolley, R. G., (1998). Is there a quantum definition of a molecule? *Journal of Mathematical Chemistry*, *23*(1–2), 3–12.
76. Yan, L., & Ollitrault, J. Y., (2014). Universal fluctuation-driven eccentricities in proton-proton, proton-nucleus, and nucleus-nucleus collisions. *Physical Review Letters*, *112*(8), 082301.
77. Yim, M. S., & Caron, F., (2006). Life cycle and management of carbon-14 from nuclear power generation. *Progress in Nuclear Energy*, *48*(1), 2–36.

78. Yokuş, A., (2021). Construction of different types of traveling wave solutions of the relativistic wave equation associated with the Schrödinger equation. *Mathematical Modelling and Numerical Simulation with Applications*, *1*(1), 24–31.
79. Zazzo, A., & Saliège, J. F., (2011). Radiocarbon dating of biological apatites: A review. *Palaeogeography, Palaeoclimatology, Palaeoecology*, *310*(1–2), 52–61.
80. Zhuang, Z., Kang, Q., Wang, D., & Li, Y., (2020). Single-atom catalysis enables long-life, high-energy lithium-sulfur batteries. *Nano-Research*, *13*, 1856–1866.

CHAPTER 2

Atomic Theory and Models

Contents

2.1. INTRODUCTION

The formulation of the first modern atomic theory by John Dalton in the early 19th century was a significant milestone. Dalton's theory presented a groundbreaking concept posing atoms as the fundamental, indivisible constituents of elements, characterized by their distinct properties. This significant development established the foundation for subsequent progress in comprehending atomic phenomena as researchers commenced investigating the characteristics and dynamics of these fundamental constituents (Mirny & Shakhnovich, 2001; Hofer et al., 2003).

Niels Bohr significantly impacted atomic theory during the early 20th century. The Bohr atomic model revolutionized the understanding of atomic structure by introducing discrete energy levels, postulating that electrons are confined to distinct orbits encircling the nucleus. The model above effectively discussed the atomic spectrum of hydrogen and stimulated the advancement of quantum mechanics, thereby revolutionizing our comprehension of the atomic domain (Bor, 1985).

Throughout the chapter, further developments in atomic study are revealed, encompassing the identification of isotopes and nuclear constituents and the fundamental principles of quantum mechanics, such as wave-particle duality and the uncertainty principle. The developments discussed above gave rise to the quantum physical model of the atom, effectively establishing a connection between classical physics and the perplexing realm of the quantum universe (Kragh, 1979; Büchler et al., 2005).

Furthermore, this chapter sheds light on the contemporary atomic theories that go beyond the domain of singular atoms. The amalgamation of quantum field theory (QFT) and the standard model of particle physics offers a comprehensive theoretical framework that facilitates the comprehension of elementary particles and their interactions. This framework aids in exploring the intricate nature of the fundamental forces that govern the universe (Fafalis et al., 2012).

The research into the atomic world is further strengthened by including key phenomena such as isomerism, Mendeleev's periodic table, and the intricate interactions among subatomic particles. This narrative explores the vast array of human intellectual contributions that have shaped our present comprehension of atoms and their far-reaching implications, ranging from the profound insights of ancient philosophers to the intricate mathematical formulations of quantum mechanics (Rico et al., 2014).

Every scientific discovery has further enhanced our understanding and admiration for the microscopic realm's intricate and aesthetically pleasing nature, catalyzing scientists to expand the frontiers of understanding and unravel the enigmatic phenomena that reside within the fundamental constituents of the universe (Sabbatucci & Salvat, 2016).

Tip

The concept of atoms dates back to the ancient Greek philosopher Democritus, who postulated the existence of tiny, indivisible particles called "atoms" over 2,500 years ago. His idea laid the foundation for the modern atomic theories we know today.

2.2. ATOMIC THEORY

Atomic theory is a fundamental concept in chemistry and physics that describes the nature and behavior of atoms, which are the building blocks of matter (Fano, 1956). Developed over centuries, atomic theory explains that:

1. **Atoms are Indivisible:** Early atomic theories proposed that atoms are the smallest, indivisible particles of matter. However, this notion was later revised as subatomic particles, like protons, neutrons, electrons, and other sub-atomic particles were discovered (Feynman, 1954).
2. **Elements Consist of One Type of Atom:** Elements, the purest form of matter, are composed of atoms with identical properties. Each element is uniquely defined by the number of protons in its nucleus, known as its atomic number.
3. **Atoms Combine to Form Compounds:** Atoms can combine in fixed ratios to create compounds. These combinations occur through chemical reactions, where atoms either share or exchange electrons to form bonds.
4. **Conservation of Mass:** The total mass of the reactants in a chemical reaction is equal to the total mass of the products. This principle, known as the law of conservation of mass, supports the idea of atoms being conserved during chemical reactions.
5. **Atomic Structure:** Atoms consist of a central nucleus composed of protons and neutrons, surrounded by electrons orbiting in energy levels or electron shells. The electrons determine an atom's chemical behavior.

6. **Isotopes:** Some elements have different forms known as isotopes, which have the same number of protons but different numbers of neutrons. Isotopes may exhibit different atomic masses and properties.

Atomic theory has evolved over time, with contributions from scientists like John Dalton, Niels Bohr, Erwin Schrödinger, and Werner Heisenberg. Today, atomic theory is an essential framework for understanding the structure and behavior of matter at the atomic and subatomic levels, influencing various scientific disciplines, from chemistry and physics to materials science and nanotechnology.

Remember

John Dalton's atomic model, proposed in the early 19th century, likened atoms to solid, indivisible spheres, much like billiard balls. This model was the first scientific attempt to describe the structure of atoms based on experimental evidence.

2.3. EARLY ATOMIC MODELS

The exploration of the composition of matter commenced many centuries ago with the emergence of early atomic models. The ancient Greeks were among the early thinkers who contemplated the essence of matter, postulating the concept of indivisible entities referred to as atoms.

Throughout history, notable developments in scientific research and empirical observations have contributed to the development of diverse atomic theories, establishing the fundamental basis for our contemporary comprehension of atoms and their impact on the structure of the cosmos. This section will examine the significant contributions made by key figures in developing early atomic models, including Democritus, Dalton, Thomson, and Rutherford. These individuals played crucial roles in laying the foundation for the subsequent atomic revolution (Desrochers et al., 2010).

2.3.1. Democritus and the Idea of Indivisible Atoms

During the 5th century BCE in ancient Greece, the philosopher Democritus proposed the concept of "atoms," which he derived from the Greek term "atomos," signifying indivisibility. Democritus speculated that these tiny entities constituted the fundamental constituents of all material substances.

The proposal suggested that atoms exhibited variations in their dimensions, configurations, and spatial organization, thereby explaining the wide-ranging characteristics observed in different substances. Although lacking empirical support, his atomic theory served as a fundamental framework, laying the groundwork for subsequent inquiries into the fundamental properties of matter (Baldes, 1978).

2.3.2. Thomson's Plum Pudding Model

J.J. Thomson, in 1904, put forth a theoretical framework for the structure of the atom, commonly referred to as the "plum pudding model" or "raisin cake model," which was developed in light of his prior identification of electrons. In this theoretical framework, the atom is conceptualized as a spherical entity with a uniform positive charge. At the same time, electrons are dispersed within this structure, like raisins embedded within plum pudding. The distribution of positive and negative charges within the atom was uniform, leading to its overall electrical neutrality (Lovisetti & Giliberti, 2022).

The plum pudding model was an initial attempt to integrate the recently identified subatomic particles within the framework of atomic structure. Nevertheless, subsequent experimental findings posed a challenge to this prevailing model, thereby facilitating the emergence of Rutherford's nuclear model (Hon & Goldstein, 2013) (Figure 2.1).

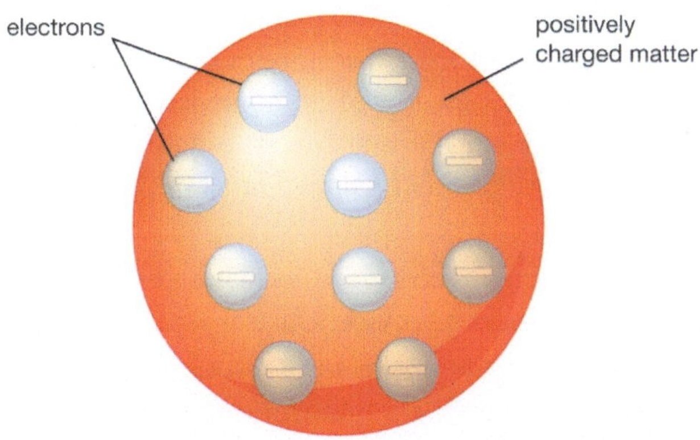

Figure 2.1. *Illustration of Thompson's plum pudding model.*

Source: Britannica, Creative Commons License.

2.3.3. Rutherford's Nuclear Model

The nuclear model of the atom was formulated in 1911 due to Ernest Rutherford's pioneering experiments involving alpha particles. The model proposed by Rutherford postulated that most of an atom's mass and positive charge were localized within a tiny and compact nucleus at its core. The electrons, which possess significantly less mass than the nucleus, revolved around the nucleus at a significant distance (Jackman et al., 1982) (Figure 2.2).

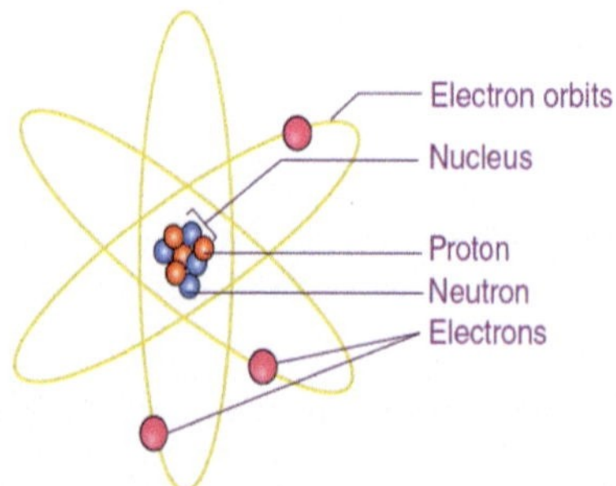

Figure 2.2. *Rutherford's atomic model.*

Source: BYJU'S, Creative Commons License.

Rutherford's model of the atom, proposed by Ernest Rutherford in 1911, included the following key postulates:

1. **Central Nucleus:** Rutherford's model suggested that most of the mass and positive charge of an atom is concentrated in a tiny, dense center called the nucleus. The nucleus contains protons, which are positively charged, and most of the atom's mass.
2. **Electron Orbits:** Electrons, carrying a negative charge, orbit the nucleus in specific paths or orbits. These orbits are at a distance from the nucleus and provide space for the electrons to move.
3. **Empty Space:** Most of the atom's volume is empty space, with the electrons occupying a relatively small portion of it while the majority of the mass is concentrated in the nucleus.
4. **Neutral Atom:** The atom as a whole is electrically neutral, meaning it has an equal number of protons (positive charges) in the nucleus and electrons (negative charges) orbiting the nucleus, resulting in no net charge.

Rutherford's model marked a significant shift in understanding the structure of the atom, especially in challenging the previously accepted idea of the "plum pudding" model proposed by J.J. Thomson. It emphasized the concentrated positive charge in the nucleus, paving the way for further developments in atomic theory (Badash, 1983; Leone et al., 2018).

2.4. DALTON'S LAW OF MULTIPLE PROPORTIONS

During the latter part of the 18th century, two laws about chemical reactions were formulated without any explicit reference to the atomic theory concept. One fundamental principle in the field of chemistry is the law of conservation of mass, which is closely linked to the contributions of Antoine Lavoisier. This law asserts that the overall mass of a chemical reaction remains unchanged, meaning that the mass of the reactants is equivalent to the mass of the resulting products.

The second principle pertained to the law of definite proportions. The law, initially formulated by Joseph Proust, a chemist from France, in the year 1797, posits that when a compound is decomposed into its constituent chemical elements, the masses of these constituents will consistently exhibit fixed weight proportions. This principle remains valid irrespective of the quantity or origin of the original substance (Niaz, 2001).

John Dalton conducted a comprehensive analysis of empirical data collected by himself and fellow scientists, identifying a recurring pattern now recognized as the law of multiple proportions.

In compounds comprising a common element, the elemental content within these compounds will exhibit variations that can be expressed as ratios of small integers. Dalton's deduction from these observations was that elements exhibit discrete and consistent units of weight when they undergo chemical reactions with each other. Dalton coined the term "atoms" by drawing upon the philosophical tradition (Murray, 1819).

2.4.1. Dalton's Atomic Theory

John Dalton, a prominent chemist and physicist from England during the early 19th century, further developed and refined the concepts initially proposed by Democritus, ultimately establishing the foundation for the first comprehensive atomic theory in modern science. The theory, which was put forth in 1808, comprised several postulates (Rocke, 2005).

- Elements consist of tiny, indivisible entities known as atoms.
- Atoms belonging to the same element exhibit uniform mass and properties, whereas atoms of distinct elements possess discernible attributes.
- Atoms tend to combine in precise and uncomplicated proportions, thereby giving rise to the formation of compounds.
- Chemical reactions entail the reorganization of atoms while the atoms themselves remain unaltered (Grossman, 2017).

The atomic theory proposed by Dalton was a groundbreaking scientific framework offering a comprehensive explanation of chemical combination principles. Moreover, it established the fundamental basis for conserving mass during chemical reactions (Peterman & Cordes, 2021).

Dalton formulated his atomic theory based on the empirical observations derived from the law of multiple proportions. One of the primary challenges faced by the theory was the determination of the comparative magnitudes of atomic weights across different elements.

The atomic weight of an element refers to the relative weight of an atom of that element in comparison to the weights of atoms belonging to other elements.

During the 19th century, Dalton and his contemporaries encountered limitations in measuring the absolute weight of atoms, specifically in grams, due to the tiny size of atoms that rendered direct measurement unfeasible with the available technologies. In contrast, the researchers measured the relative atomic masses of different elements compared to hydrogen, which chemists recognized during Dalton's era as the element with the lowest atomic mass (Thackray, 1966) (Figure 2.3).

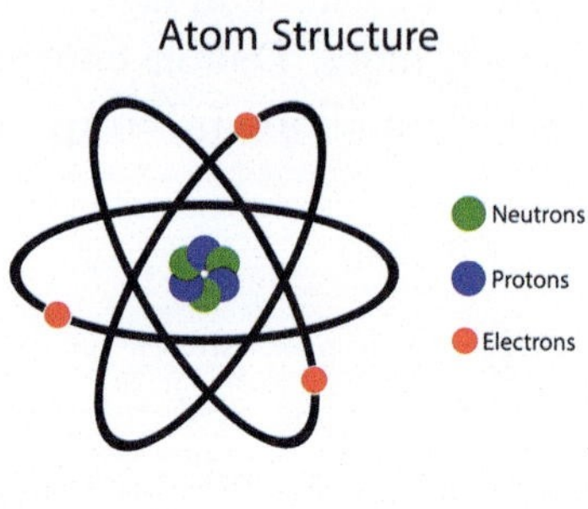

Figure 2.3. *Atomic structure of an atom Dalton's atomic theory.*

Source: YClass, Creative Commons License.

Dalton made estimations of atomic weights based on the ratios of masses they combined, using the weight of the hydrogen atom as a conventional unit. Nevertheless, Dalton failed to recognize that certain elements naturally exist as molecules in their pure form, such as pure oxygen as O_2. Furthermore, he erroneously believed that the most basic compound formed between any two elements invariably consists of a single atom of each element. Consequently, he incorrectly postulated that the chemical formula for water is HO rather than the correct representation of H_2O. Furthermore, the constraints imposed by his experimental equipment compromised the validity of his findings.

In 1803, he theorized that the weight of oxygen atoms was approximately 5.5 times greater than that of hydrogen atoms. This conclusion was drawn based on the observation that there existed a ratio of 5.5 grams of oxygen to every 1 gram of hydrogen in water, leading to the belief that the chemical formula for water was HO. Based on improved data, it was determined in 1806 that the atomic weight of oxygen should be revised from 5.5 to 7. This revised value was consistently upheld by him throughout their lifetime. At present, others have determined through more accurate measurements that the relative weight of an oxygen atom is 8 about hydrogen, assuming Dalton's formula for the water molecule (HO), or 16 if the modern water formula (H_2O) is assumed (Kedrov, 1949).

Amedeo Avogadro rectified the inherent flaw in Dalton's theory in 1811, in principle. Avogadro postulated that under identical temperature and pressure conditions, equal volumes of different gases would contain an equal quantity of molecules, thereby implying that the mass of a gas's particles does not impact the volume it occupies. Avogadro's hypothesis, commonly called Avogadro's law, offers a means of determining the comparative masses of gaseous molecules. According to this hypothesis, if correct, the relative densities of gases can directly signify the relative masses of the constituent particles within them. This line of reasoning gave rise to a subsequent conjecture (Pratt, 2010): the constituent particles of specific elemental gases were not individual atoms but rather molecular entities composed of two atoms each; furthermore, these molecular entities frequently underwent division into two separate entities upon undergoing chemical combination.

The observation that combining two liters of hydrogen with one liter of oxygen yields two liters of water vapor under constant pressure and temperature implies that a single oxygen molecule undergoes dissociation to generate two water molecules. This also implies that the chemical composition of the water molecule is H_2O. Avogadro's contributions enabled him to provide more precise approximations of the atomic mass of oxygen and other elements while establishing a clear differentiation between molecules and atoms. The term "elementary molecules" was employed by Avogadro to refer to what is presently known as atoms. In contrast, the term "compound molecules" was used by Avogadro to denote what is presently known as molecules (Hartley, 1967).

2.4.2. Opposition to Atomic Theory

The acceptance of Dalton's atomic theory was not immediate among the scientific community. One issue that arose was the absence of consistent terminology. The term "atom" originally indicated indivisibility; however, Dalton introduced a different definition, considering an atom as the fundamental constituent of any substance. Consequently, this definition allowed for the division of "compound atoms" like carbon dioxide, in contrast to "elementary atoms." Additional researchers employed their naming system, thereby contributing to the overall state of perplexity. One example of this can be seen in the work of J. J. Berzelius, who employed the phrase "organic atoms" to denote particles comprising three or more elements, as he believed these entities were exclusively present in organic compounds (Sylvester, 1878).

During the 19th century, scientists lacked the means to observe atoms directly. The existence of atoms was inferred using indirect observations, exemplified by Dalton's law of multiple proportions. Certain scientists, particularly those adhering to the positivist school of thought, posited that scientists must not attempt to infer the underlying nature of the cosmos but rather confine their efforts to organizing and categorizing the observable patterns they can directly perceive.

The proponents of anti-atomism posited that although atoms may serve as a valuable conceptual tool for forecasting elemental interactions, they do not accurately correspond to tangible phenomena (Roller, 1981).

These scientists were occasionally referred to as "equivalents" due to their preference for the theory of equivalent weights, which can be seen as an extension of Proust's law of definite proportions. As an illustration, the chemical reaction between 1 gram of hydrogen and 8 grams of oxygen forms 9 grams of water. Consequently, the equivalent weight of oxygen can be determined as 8 grams. The standpoint above was ultimately invalidated by two significant advancements that transpired in the latter half of the 19th century: the emergence of the periodic table and the revelation that molecules possess an inherent structure that governs their characteristics.

The universality of Dalton's law of multiple proportions was limited in the context of organic substances. For example, oleic acid exhibits a hydrogen-to-carbon ratio of 34 g to 216 g, while methane demonstrates a hydrogen-to-carbon ratio of 72 g to 216 g. The numbers 34 and 72 exhibit a ratio of 17:36, which does not conform to a ratio of small integers. Carbon-based compounds are presently understood to have the capacity to form significantly larger molecules than those formed by other elements. The chemical formula of oleic acid is $C_{18}H_{34}O_{2,}$ while CH_4 represents methane (Furley, 1987).

2.5. BOHR'S MODEL OF ATOM

During the early 20th century, the field of physics explored the enigmatic domain of atomic phenomena. During this time, Niels Bohr introduced a groundbreaking conceptual framework that revolutionized our comprehension of the atom. The model proposed by Bohr in 1913 marked a substantial departure from classical physics and established the fundamental principles of quantum mechanics. By integrating principles from classical physics and quantum theory, Bohr effectively discussed the atomic spectrum of hydrogen, thereby unveiling novel understandings of the dynamics of electrons and their discrete energy states within atomic structures. This section discusses the fundamental principles, mathematical foundations, and historical significance of Bohr's atomic model (Sigmund & Haagerup, 1986).

2.5.1. Background and Need for a New Model

Classical physics encountered a significant challenge during the late 19th and early 20th centuries when attempting to explain atomic systems, as it relied on Newtonian mechanics and electromagnetism as its foundational principles. Based on classical theories, it is postulated that electrons in orbit around the nucleus undergo a continuous energy loss, ultimately leading to their collapse into the nucleus. This phenomenon implies an inherent instability within atoms. Nevertheless, the empirical data obtained from atomic spectra challenged this hypothesis, necessitating the development of a novel framework capable of harmonizing the observed phenomena with the fundamental tenets of classical physics.

2.5.2. Bohr's Postulates and the Hydrogen Spectrum

Bohr formulated a theoretical framework that encompassed several postulates, thereby establishing the fundamental principles of his model:

1. The presence of electrons within an atom is restricted to specific discrete energy levels or stationary orbits. The orbits above exhibit stability and do not undergo energy emission, in contrast to the continuous energy dissipation anticipated by classical physics.
2. The concept of quantization of angular momentum was introduced by Bohr, who postulated that the angular momentum of an electron in its orbit is restricted to discrete, quantized values. The angular momentum (L) can be expressed as the multiplication of the mass of the electron (m), the velocity of its orbit (v), and the radius of said orbit (r). The mathematical representation of the given concept is denoted as:

$$L = mvr = n\hbar \qquad (2.1)$$

where; 'n' represents an integer known as the principal quantum number; and ħ (h-bar) symbolizes the reduced Planck constant (h/2π) (Makhmudov & Khudayberganov, 2023).

3. The phenomenon of energy transitions involves the transfer of electrons between different energy levels through the absorption or emission of discrete units of energy known as photons. Conversely, the phenomenon of an electron transitioning from a higher to a lower energy level through the emission of a photon is observed (Eremin et al., 2020) (Figure 2.4).

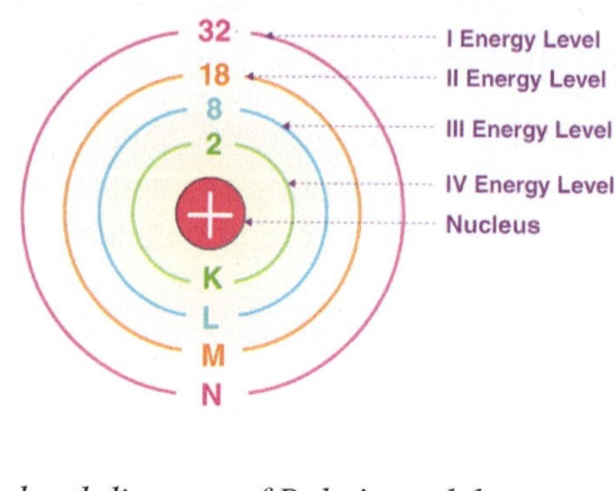

Figure 2.4. *Energy level diagram of Bohr's model.*

Source: BYJU'S, Creative Commons.

2.5.3. The Mathematics of Bohr's Model

Bohr's model introduced quantized energy levels for electrons, given by the formula.

$$E_n = -\frac{(Z^2 \times R_H)}{n^2} \tag{2.2}$$

The energy of an electron in the nth energy level, denoted as E_n, is determined by the atomic number (Z) of the atom, the Rydberg constant (R_H), and the principal quantum number (n).

The formula demonstrates that as the variable n increases, energy levels converge, indicating the discrete nature of these levels. The electron's energy experiences a decrease in magnitude (i.e., a greater degree of binding to the nucleus) as the principal quantum number, n, increases. The electron with the highest degree of binding is found in the lowest energy level, corresponding to n = 1.

2.5.4. Limitations and Challenges of the Bohr Model

Although Bohr's model represented a notable progression in the field, it was not without its constraints.

The model was restricted to hydrogen-like atoms, characterized by the presence of a solitary electron, and failed to accommodate atoms of greater complexity that possess multiple electrons. Furthermore, Bohr's model could not provide a comprehensive explanation for the fundamental factors that account for the quantization of energy levels and the stability of electrons within these orbits (Turner & Rowe, 2005).

2.5.5. Bohr's Model and the Quantum Revolution

Although Bohr's model has limitations, it is considered a pivotal moment in physics.

The concept of quantization was introduced and served as a foundational basis for subsequent advancements in quantum mechanics. The model effectively discussed the atomic spectrum of hydrogen, yielding precise prognostications for the wavelengths of the spectral lines. The model proposed by Bohr served as a crucial link between classical and quantum physics, illustrating the indispensability of quantum principles in comprehending atomic behavior.

2.5.6. Legacy and Significance

The Bohr model of the atom played a pivotal role in the evolution toward the formulation of more advanced quantum mechanical models. The discovery above catalyzed subsequent scholars to investigate the wave-particle duality exhibited by electrons, the principle of uncertainty, and the Schrödinger wave equation. These inquiries ultimately culminated in developing the contemporary quantum mechanical framework for understanding the atom.

The principles postulated by Bohr remain foundational in quantum mechanics and are indispensable in our comprehension of the dynamics exhibited by subatomic entities. The contributions made by his research laid the foundation for the investigation of atomic and subatomic physics, facilitating our comprehension of the complexities inherent in the quantum field and its significant implications for our comprehension of the cosmos.

The model of the atom proposed by Niels Bohr represents a significant milestone in the field of physics, as it revolutionized our understanding of the atomic domain. By introducing discrete energy levels and harmonizing classical physics with quantum concepts, Bohr effectively discussed the atomic spectrum of hydrogen and laid the foundation for the advancement of quantum mechanics. Although the Bohr model possessed certain limitations, its impact on the development of our comprehension of atomic behavior and its role in stimulating subsequent investigations in the field of quantum physics should not be underestimated. The model proposed by Bohr continues to be a fundamental aspect in the interpretation of the fundamental aspects of the quantum field and its significant influence on the formation and development of the cosmos (Shaik, 2007).

2.6. ISOMERISM

Researchers quickly identified substances with identical elemental composition ratios but with distinct characteristics. In 1827, Friedrich Wöhler made a significant discovery regarding silver fulminate and cyanate composition. He found that both substances consist of 107 parts of silver, 12 parts of carbon, 14 parts of nitrogen, and 12 parts of oxygen. These compounds are now recognized by the chemical formula $AgCNO$. Wöhler's findings also revealed that urea and ammonium cyanate possess identical chemical compositions, with their respective formulas now recognized as CH_4N_2O, despite exhibiting distinct properties. The term isomerism was introduced by Jöns Jacob Berzelius in the year 1830 to describe this particular phenomenon. The proposition that isomerism arises from the varying configurations of identical quantities and types of atoms, leading to the formation of distinct substances, was widely embraced by chemists during the 1830s and subsequent years. The proliferation of isomers experienced significant growth following the advancement of organic chemistry, particularly with the emergence of atomic valence and structural theory in the 1860s. Consider, as an illustrative instance, the compound pentane (C_5H_{12}). Based on valence and molecular structure principles, it is postulated that the pentane molecule can exist in three distinct atomic arrangements. Consequently, three distinct substances possess identical compositional elements as pentane yet exhibit divergent properties (Sansone et al., 2013) (Figure 2.5).

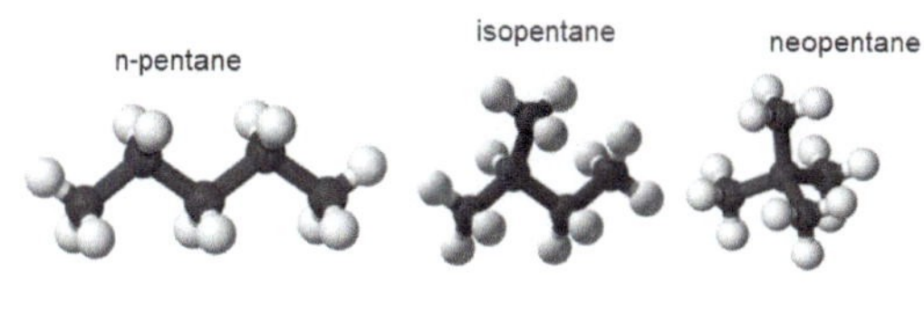

Figure 2.5. *Three isomers of pentane (C_5H_{12}): Black spheres signify carbon atoms while white spheres signify hydrogen atoms.*

Source: Andrew, Creative Commons License.

Louis Pasteur proposed in 1860 that isomers, despite having the same chemical composition, could possess distinct spatial arrangements of their atoms. In the year 1874, Jacobus Henricus Van't Hoff put forth the proposition that the carbon atom engages in bonding with other atoms in a tetrahedral configuration. Based on this hypothesis, the author could discuss instances of isomerism in which the molecules exhibit an apparently identical fundamental structure. The sole distinction between these two molecules lies in their respective three-dimensional spatial arrangements, akin to a pair of indistinguishable left and right hands or two identical spirals that coil in opposite directions, one clockwise and the other counterclockwise (Sarathy et al., 2012).

Did You Know?

Bohr's model of the atom introduced the concept of quantized energy levels, where electrons can only exist in specific orbits around the nucleus. This model successfully explained the atomic spectrum of hydrogen but was later replaced by more advanced quantum mechanical models.

2.7. MENDELEEV'S PERIODIC TABLE

Dmitrii Mendeleev observed a distinct periodicity among the elements when he sequentially organized them based on their atomic weights. As an example, it is observed that the second element, lithium, exhibits comparable properties to the ninth element, sodium, and the 16th element, potassium, thereby forming a periodicity of seven elements. Similarly, beryllium, magnesium, and calcium exhibited resemblances, with each element being separated by a distance of seven positions on Mendeleev's periodic table (eight positions on the contemporary table) (Laing, 2008). By utilizing these patterns, Mendeleev made anticipations regarding the presence and characteristics of novel elements, subsequently verified to occur naturally as scandium, gallium, and germanium. Furthermore, the periodic table can forecast the number of atoms from different elements with which an atom can form chemical bonds. For instance, germanium and carbon, being situated in the same group on the periodic table, can combine with two oxygen atoms each, resulting in the formation of GeO_2 and CO_2. Mendeleev's discovery of these patterns substantiated the hypothesis positing the atomic nature of matter, as it demonstrated the feasibility of classifying elements based on their respective atomic weights. Introducing a novel element into the middle of a given period would disrupt the parallelism between said period and the subsequent one, thereby contravening Dalton's law of multiple proportions (Karol, 2002).

2.7.1. Brownian Motion

In 1827, the esteemed British botanist Robert Brown observed the behavior of dust particles within pollen grains suspended in water, wherein he noted their incessant and seemingly purposeless movement.

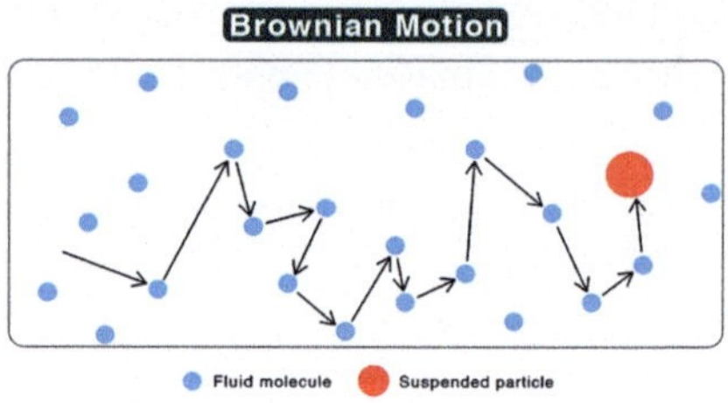

Figure 2.6. *Illustration of Brownian motion.*

Source: Science Facts, Creative Commons License.

In 1905, Albert Einstein postulated that the phenomenon known as Brownian motion could be attributed to the constant collisions between water molecules and grains and subsequently formulated a mathematical framework to discuss this phenomenon. The experimental validation of this model was conducted in 1908 by Jean Perrin, a French physicist. Perrin utilized Einstein's equations to ascertain the dimensions of atoms (Figure 2.6).

2.7.2. Statistical Mechanics

The postulation of atomic existence was necessary to introduce the Ideal Gas Law and statistical physics. In 1738, Daniel Bernoulli, a physicist and mathematician hailing from Switzerland, proposed that the origins of gas pressure and heat can be attributed to the fundamental movement of molecules.

In the year 1860, James Clerk Maxwell, a prominent advocate of atomism, became the pioneer in applying statistical mechanics within the field of physics. Ludwig Boltzmann and Rudolf Clausius further developed the field of thermodynamics, specifically focusing on the study of gases and the laws governing them, emphasizing the second law, which pertains to the concept of entropy.

During the 1870s, Josiah Willard Gibbs made significant advancements in the field of thermodynamics by expanding the principles of entropy and introducing the concept of "statistical mechanics."

Einstein subsequently rediscovered Gibbs' laws through independent inquiry, as their original publication was limited to a relatively unknown American journal. Einstein subsequently expressed that had he possessed knowledge of Gibbs' contributions, he would have refrained from publishing those papers altogether, instead focusing solely on addressing specific and unique aspects. The fundamental principles of statistical mechanics, encompassing the laws governing heat, gas behavior, and entropy, were predicated upon the essential assumption of the existence of atoms (Amaral et al., 2001).

2.7.3. Discovery of Subatomic Particles

During the early 20th century, significant experimental breakthroughs were made, unveiling the existence of subatomic particles within atoms.

2.7.3.1. Electrons

The discovery of electrons can be attributed to J.J. Thomson's cathode ray tube (CRT) experiments conducted in 1897. The researcher observed that cathode rays, which were subsequently recognized as streams of electrons, exhibited deflection when subjected to electric and magnetic fields. This deflection indicated the existence of negatively charged particles within atoms. Thomson's research yielded the initial empirical substantiation for the presence of subatomic particles (Kibler, 2007).

2.7.3.2. Protons

The renowned gold foil experiment was conducted in 1919 by Ernest Rutherford, Hans Geiger, and Ernest Marsden. The researchers subjected a slender gold foil to a bombardment of alpha particles and recorded the resulting scattering patterns. Remarkably, a subset of alpha particles exhibited significant deflection at considerable angles, prompting Rutherford to postulate the nuclear model of the atom. The hypothesis posited that most of an atom's mass and positive charge were localized within a compact and dense area referred to as the nucleus, housing the subatomic particles known as protons.

2.7.3.3. Neutrons

Neutrons were first discovered by James Chadwick in 1932. These subatomic particles possess no charge yet exhibit a mass comparable to a proton. This discovery provided a comprehensive understanding of the atomic nucleus, wherein most of the atom's mass comprises protons and neutrons.

The prevailing belief regarding the fundamental unit of matter was that atoms represented the ultimate division level until 1897, when J. J. Thomson's investigations into cathode rays led to the discovery of the electron (Oganessian, 2019).

A Crookes tube refers to a hermetically sealed glass vessel that contains a vacuum and is equipped with two electrodes positioned at a distance. Applying a voltage across the electrodes generates cathode rays, which in turn produce a luminous region upon striking the glass surface located at the opposite end of the tube.

Thomson's experimental findings revealed that the rays exhibited deflection when subjected to an electric field, expanding the knowledge of their deflection solely by magnetic fields. The researcher concluded that these rays, instead of being classified as a type of light, consisted of extremely lightweight particles with a negative charge (Van Tiggelen & Lykknes, 2019). Thomson referred to these entities as "corpuscles;" however, other scientists adopted the term "electrons" to designate them, by a proposal by George Johnstone Stoney in 1894 to designate the fundamental unit of electrical charge. The researcher conducted measurements of the mass-to-charge ratio and observed that it exhibited a value 1800 times lesser than that of hydrogen, which is acknowledged as the smallest atom. These corpuscles exhibited unique characteristics that distinguished them from any previously identified particle (Selco et al., 2013). Thomson postulated the divisibility of atoms and proposed that corpuscles served as their fundamental constituents. He posited that the corpuscles were uniformly dispersed within a sea of positive charge to discuss the atom's overall neutral charge. The model above acquired the appellation of the plum pudding model due to the conceptualization of electrons being nestled within a positively charged medium, analogous to the arrangement of fruit particles within a desiccated-fruit pudding. However, Thomson postulated that the electrons had mobility within the atomic structure (Ghosh, 2020) (Figure 2.7).

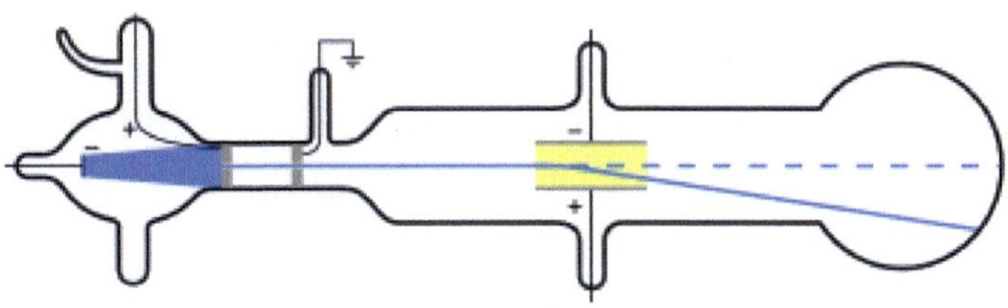

Figure 2.7. *Illustration of a cathode ray tube.*

Source: Raymond, Creative Commons License.

Remember

Heisenberg's Uncertainty Principle states that it is impossible to simultaneously measure both the position and momentum of a subatomic particle with absolute precision. The more accurately one property is measured, the less accurately the other can be known.

2.7.4. Discovery of the Nucleus

The plum pudding model proposed by Thomson was invalidated in 1909 through the findings of Ernest Rutherford, a former student of Thomson.

Rutherford's research revealed that most of an atom's mass and positive charge are concentrated within a minute portion of its overall volume, which he postulated to be located in the atom's central region (Andriiko & Lunk, 2018).

Ernest Rutherford and his colleagues Hans Geiger and Ernest Marsden developed skepticism towards the Thomson model as they encountered challenges during their attempts to construct an apparatus for determining the charge-to-mass ratio of alpha particles. These particles, which possess positive charges and are emitted by radioactive substances like radium, posed complications in their experimental studies. The scattering of alpha particles by the air in the detection chamber resulted in a lack of reliability in the measurements obtained. Thomson faced a comparable challenge in his investigation of cathode rays, which he effectively resolved by establishing a near-total vacuum within his apparatus. Rutherford did not anticipate encountering this particular issue due to the significantly greater mass of alpha particles than electrons. Based on Thomson's atomic model, it is posited that the atom's positive charge is insufficiently concentrated to generate an electric field of significant strength capable of deflecting an alpha particle. Additionally, due to the comparatively lighter mass of electrons, they are expected to be easily displaced by the significantly heavier alpha particles. However, the occurrence of scattering prompted Rutherford and his colleagues to undertake a meticulous investigation of this phenomenon (Schwerdtfeger et al., 2020).

During the period from 1908 to 1913, a series of experiments were conducted by Rutherford and his colleagues. These experiments involved the bombardment of thin metal foils with alpha particles. Alpha particles were observed to undergo deflection at angles exceeding 90°. To explain, Rutherford postulated that the atom's positive charge is not uniformly dispersed within the entirety of the atom's volume, as Thomson had previously posited. Instead, Rutherford proposed that this positive charge is highly localized within a tiny nucleus at the atom's central region. The observed deflection of alpha particles can be attributed to an electric field, which is generated by a high charge concentration. Rutherford's proposed model is occasionally called the "planetary model." In 1904, Hantaro Nagaoka was credited by Rutherford as being the originator of the concept of a planetary atom. The proposition of planetary models, including Joseph Larmor's model, dates back to 1897.

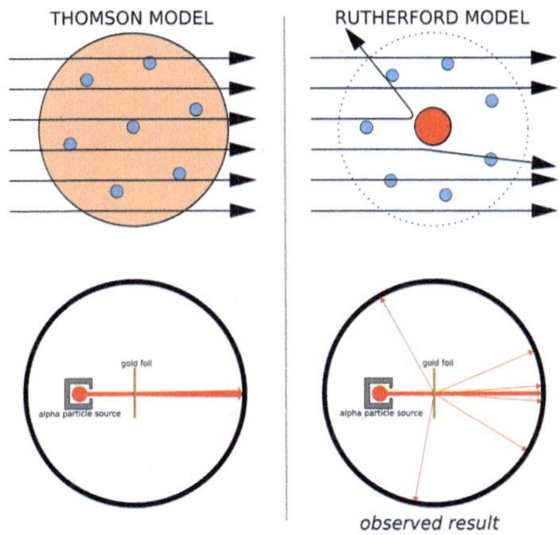

Figure 2.8. *Left: Alpha particles passing through the plum pudding model of the atom with negligible deflection. Right: Observed results; a small portion of the particles deflected by the concentrated positive charge of the nucleus.*

Source: Charles, Creative Commons License.

The earliest known solar system model can be traced back to an unpublished note by Ludwig August Colding in 1854. Colding proposed that atoms resemble planetary systems, exhibiting rotational motion and giving rise to magnetic polarity (Gol'Danskii & Avakian, 1970) (Figure 2.8).

2.7.5. First Steps Toward a Quantum Physical Model of the Atom

The planetary model of the atom exhibited two notable deficiencies. One notable distinction is that electrons, in contrast to planets revolving around a star, possess an electric charge. The emission of electromagnetic waves by an electric charge undergoing acceleration is a well-established phenomenon in classical electromagnetism, as described by the Larmor formula. According to classical electromagnetic theory, a charged particle in orbit is expected to undergo a gradual dissipation of energy, resulting in a progressive inward spiral toward the nucleus. Eventually, this process would lead to a collision between the orbiting charge and the nucleus, occurring within a brief temporal interval. One additional issue pertained to the inadequacy of the planetary model in explaining the prominently concentrated emission and absorption spectra exhibited by observed atoms (Labuhn et al., 2016) (Figure 2.9).

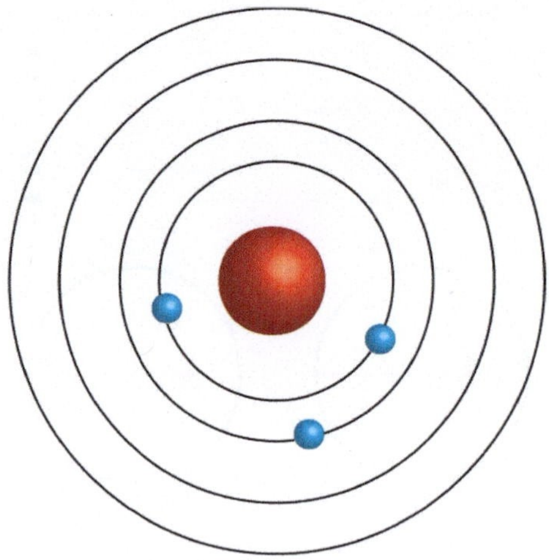

Figure 2.9. *The Bohr model of the atom.*

Source: Charles, Creative Commons License.

Physics underwent a significant transformation in the early 20th century due to the advent of quantum theory. This groundbreaking theory, proposed by Max Planck and Albert Einstein, posited that the emission or absorption of light energy occurs in discrete units referred to as quanta, with each unit being termed a quantum. As a result, a sequence of quantum atomic models emerged, including Arthur Erich Haas' quantum model in 1910 and John William Nicholson's quantum atomic model in 1912, which introduced the quantization of angular momentum as $h/2\pi$. In 1913, Niels Bohr integrated this concept into his Bohr atom model (Kuzmich et al., 2003).

According to this model, electrons were restricted to specific circular orbits around the nucleus, each characterized by a fixed angular momentum and energy. The distance between the electron and the nucleus, their radii, is directly proportional to the electron's energy. According to this theoretical framework, the possibility of an electron spiraling into the nucleus is precluded due to its inability to undergo a continuous energy loss. Instead, the electron is constrained to undergo instantaneous transitions, known as "quantum leaps," between discrete energy levels. During this phenomenon, the emission or absorption of light occurred at a frequency that was directly proportional to the alteration in energy. This relationship between energy change and light emission/absorption resulted in the manifestation of discrete spectra (Steffens et al., 2015).

The model proposed by Bohr exhibited imperfections. The model's predictive capabilities were limited to the spectral lines of hydrogen, excluding those of multielectron atoms. Furthermore, it was found that Bohr's model could not fully account for the hydrogen spectrum, as advancements in spectrographic technology revealed that the application of a magnetic field resulted in the multiplication of spectral lines. Bohr's model could not explain this phenomenon. In 1916, Arnold Sommerfeld introduced elliptical orbits as an extension to the Bohr model to account for the presence of additional emission lines. However, this modification rendered the model considerably intricate to apply, and it remained inadequate in explaining the behavior of more intricate atomic systems (Westermayr & Marquetand, 2020).

2.8. DISCOVERY OF ISOTOPES

In 1913, a radiochemist, Frederick Soddy, made a significant observation during his investigation of the products resulting from radioactive decay. He found evidence suggesting the existence of multiple variations of certain elements. The terminology "isotope" was introduced by Margaret Todd to designate these distinct forms (Lal, 1988).

In the year above, J. J. Thomson performed an experimental investigation involving the passage of neon ions through magnetic and electric fields, resulting in their collision with a photographic plate positioned at the opposite end. The observer noted the presence of two luminous areas on the plate, indicating the occurrence of two distinct paths of deflection. Thomson's conclusion was based on the observation that certain neon ions exhibited a distinct mass variation. The explanation of the varying mass of atoms was subsequently provided by identifying neutrons in 1932. It was established that all atoms of a specific element possess identical protons, whereas distinct isotopes exhibit dissimilar quantities of neutrons (Eaton, 2020).

2.9. DISCOVERY OF NUCLEAR PARTICLES

In 1917, Rutherford conducted an experiment in which he subjected nitrogen gas to alpha particles and observed the emission of hydrogen nuclei from the gas. Rutherford could identify these hydrogen nuclei as he had previously observed them when bombarding hydrogen with alpha particles and examining the resulting products. Rutherford concluded that the hydrogen nuclei originated from within the nuclei of the nitrogen atoms, essentially resulting in the splitting of a nitrogen atom (Dreyfuss et al., 1988).

Based on his personal research and the research conducted by his students Bohr and Henry Moseley, Rutherford knew that the positive charge of any given atom could consistently be expressed as a whole number multiple of the charge carried by a hydrogen nucleus. The observation above, in conjunction with the fact that the atomic mass of numerous elements is approximately equal to a whole number of multiple hydrogen atoms, which were conventionally regarded as the lightest particles, prompted him to deduce that hydrogen nuclei were indivisible entities and fundamental components of all atomic nuclei. The particles were designated as protons by him. Additional investigations conducted by Rutherford revealed that the nuclear mass of most atoms surpassed the number of protons they contained. Rutherford postulated that this excess mass could be attributed to neutrally charged particles, referred to as "neutrons," provisionally (Acevedo et al., 2022).

In 1928, Walter Bothe observed that when alpha particles were directed toward beryllium, they emitted radiation with high penetrating ability and electrical neutrality. Subsequent findings revealed that the radiation above can dislodge hydrogen atoms from paraffin wax. At first, it was hypothesized that the observed phenomenon was caused by high-energy gamma radiation, as this type of radiation had been observed to have a comparable impact on electrons in metallic substances. However, James Chadwick discovered that the ionization level was excessively pronounced to be attributed to electromagnetic radiation if energy and momentum conservation principles were upheld during the interaction.

Did You Know?

Dmitri Mendeleev's periodic table, organized based on the atomic masses of elements, not only organized known elements but also predicted the existence and properties of undiscovered elements. Many elements were later found to fit the positions Mendeleev had predicted.

In 1932, Chadwick conducted an experiment wherein he subjected different elements, including hydrogen and nitrogen, to an enigmatic form of radiation known as "beryllium radiation." By measuring the energies exhibited by the charged particles that recoiled from this radiation, Chadwick inferred that the radiation consisted of electrically neutral particles. These particles were found to possess a mass that was not negligible, like that of gamma rays, but comparable to that of a proton. Chadwick subsequently attributed these particles to Rutherford's neutrons. Chadwick was awarded the Nobel Prize in 1935 for discovering neutrons (Giunta & Mainz, 2020).

2.10. QUANTUM PHYSICAL MODELS OF ATOM

In 1924, Louis de Broglie proposed that particles in motion, specifically subatomic particles like electrons, possess a certain extent of wave-like characteristics. Erwin Schrödinger, driven by a profound interest in this concept, investigated the possibility of explaining the motion of an electron within an atom by considering it as a wave phenomenon instead of a particle.

The equation formulated by Erwin Schrödinger in 1926 describes the electron in terms of a wave function, as opposed to a localized point particle. This approach successfully anticipated numerous spectral phenomena not adequately accounted for by Bohr's model. Despite the mathematical convenience of this concept, it encountered challenges in terms of visualization and resistance. Max Born, a prominent critic, put forth an alternative viewpoint suggesting that the wave function formulated by Schrödinger does not depict the physical extension of an electron, akin to a charge distribution in classical electromagnetism. Instead, Born proposed that the wave function represents the likelihood of finding an electron at a specific location upon measurement. The concept of wave-particle duality emerged to reconcile the seemingly contradictory characteristics of electrons and other subatomic entities, which exhibit wave-like and particle-like behaviors. The prominence of either aspect, wave-like or particle-like, is contingent upon the specific circumstances (Mills, 2001) (Figure 2.10).

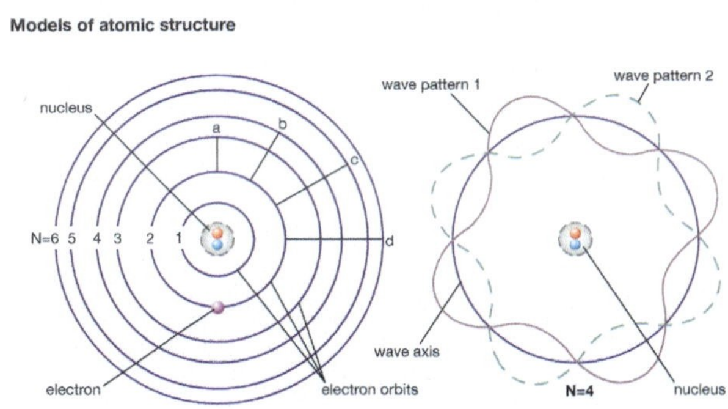

Figure 2.10. *Quantum model of the atom.*

Source: Britannica, Creative Commons License.

One implication of characterizing electrons as waveforms is the mathematical impossibility of simultaneously determining the position and momentum of an electron. The concept was subsequently recognized as the Heisenberg uncertainty principle, named after Werner Heisenberg, a prominent theoretical physicist who initially formulated this principle in 1927. Heisenberg analyzed a thought experiment involving the simultaneous measurement of an electron's position and momentum. Nevertheless, Heisenberg failed to provide precise mathematical definitions explaining the concept of "uncertainty" in these measurements (Primas, 1975). The formulation of the position-momentum uncertainty principle in mathematics is credited to Earle Hesse Kennard, Wolfgang Pauli, and Hermann Weyl. This discovery rendered Bohr's model, characterized by its orderly and distinctly delineated circular orbits, invalid. The contemporary atomic model discusses the spatial distribution of electrons within an atom by employing the concept of probability. The electron has the potential to occupy any distance from the nucleus. However, its energy level and angular momentum influence its likelihood of being found in specific regions around the nucleus. This distribution of electron probability is commonly known as its atomic orbital. The orbitals exhibit diverse shapes, including but not limited to spheres, dumbbells, and tori, with the nucleus positioned at its center. The determination of atomic orbital shapes is achieved through the resolution of the Schrödinger equation. It is worth noting that the availability of analytical solutions for the Schrödinger equation is limited to a small number of relatively uncomplicated model Hamiltonians, such as the hydrogen atom and the dihydrogen cation. Despite its minimal composition of two electrons, the helium atom has proven to be resistant to comprehensive analytical approaches (Tosoni et al., 2023) (Figure 2.11).

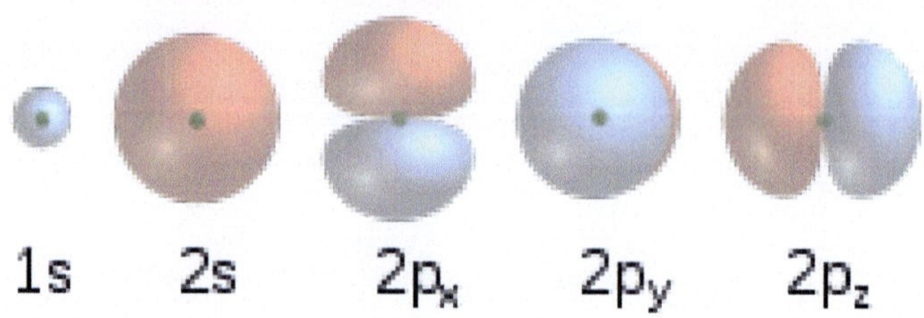

Figure 2.11. *The five filled atomic orbitals of a neon atom separated and arranged in order of increasing energy from left to right, with the last three orbitals being equal in energy.*

Source: Louis, Creative Commons License.

2.11. MODERN ATOMIC THEORIES

Throughout the 20th century, a significant paradigm shift occurred in our comprehension of atomic and subatomic phenomena. The emergence of quantum mechanics marked a significant departure from classical physics, paving the way for advancing contemporary atomic theories (Braune, 2010). The theories above, grounded in the fundamental principles of quantum mechanics, offer a thorough and precise account of the dynamics exhibited by atoms and subatomic particles. This investigation will examine contemporary atomic theories, focusing on fundamental concepts, including wave-particle duality, quantum numbers, electron orbitals, and the role of quantum field theory (QFT) in shaping our comprehension of the atomic domain (Chalmers, 1998).

2.11.1. Quantum Mechanics and Wave-Particle Duality

The wave-particle duality principle is considered a fundamental concept in contemporary atomic theories. The statement posits that particles, specifically electrons, demonstrate characteristics of both waves and particles. The initial introduction of this concept can be attributed to Louis de Broglie in 1924, wherein he postulated that particles, specifically electrons, possess wave-like characteristics interconnected with their momentum (Cheong & Song, 2014). The phenomenon of wave-particle duality in electrons was subsequently validated through empirical evidence in the renowned electron diffraction experiments. In these experiments, electrons displayed interference patterns similar to waves as they traversed a crystal lattice. The discovery above challenged the conventional understanding of particles as entities with definite positions, establishing the groundwork for advancing quantum mechanics (Hendry, 1980).

2.11.2. Quantum Numbers and Electron Orbitals

The quantum mechanical model of the atom encompasses the description of electrons through a collection of quantum numbers, which define their energy levels, angular momentum, and orientation within an atomic system. The four quantum numbers encompass (Hunt et al., 2017):

1. **Principal Quantum Number (n):** The term "orbital" refers to the specific energy level or electron shell that accommodates an electron. Greater values of n are associated with increased energy levels located at greater distances from the atomic nucleus (Moore, 1999).

2. **Azimuthal Quantum Number (l):** It determines the shape of the electron's orbital. Different values of l correspond to different subshells: s (l=0), p (l=1), d (l=2), and f (l=3) (Fakhri & Hashemzadeh, 2009).
3. **Magnetic Quantum Number (m_l):** It describes the orientation of the orbital in space relative to an external magnetic field (Willett et al., 1987).
4. **Spin Quantum Number (m_s):** It specifies the intrinsic spin of the electron, either +1/2 (spin-up) or –1/2 (spin-down).

These quantum numbers play a crucial role in determining the arrangement of electrons within atoms and their chemical behavior and interactions (Mátyus & Reiher, 2012).

2.11.3. Electron Configurations and the Aufbau Principle

The electron configuration of an atom is described by its arrangement of electrons using quantum numbers. The "Aufbau Principle," a fundamental principle in quantum mechanics, determines the electron configuration. According to this principle, electrons are arranged in an atom by filling the lowest energy levels before occupying higher ones (Guo et al., 2019).

The electron configurations offer a systematic approach to comprehending the arrangement of electrons within atoms and their allocation among various orbitals and subshells. Understanding this information is of utmost importance when explaining the characteristics and responsiveness of elements and comprehending the organization of elements within the periodic table (Metz et al., 2021).

2.11.4. Quantum Field Theory (QFT) and Particle Physics

The principles of special relativity were integrated into quantum mechanics, resulting in the advancement of QFT. QFT is a theoretical framework in which particles are conceptualized as quantized fields. The dynamics of these particles are discussed using fundamental forces, which are mediated by exchange particles (Wallace, 2011).

The Standard Model of particle physics is a fundamental element within contemporary atomic theories. The unification of three fundamental forces: electromagnetism, the weak nuclear force, and the strong nuclear force (Dunne, 2014).

2.11.5. Applications and Advancements

Contemporary atomic theories possess noteworthy implications across diverse disciplines, encompassing chemistry, materials science, and particle physics.

These concepts serve as the fundamental basis for comprehending chemical bonding, molecular structures, and the characteristics of matter on both the atomic and subatomic scales. Moreover, the foundational principles of contemporary atomic theories serve as the basis for advancing sophisticated technologies, including semiconductor devices and quantum computing (Shao et al., 2006).

2.11.6. Future Directions

Although contemporary atomic theories have yielded significant insights into the behavior of atoms and subatomic particles, there remain unresolved inquiries and uncharted domains. Researchers persist in examining subjects such as dark matter and energy, the amalgamation of all fundamental forces, and the essence of black holes in pursuit of a more all-encompassing comprehension of the cosmos.

The advent of contemporary atomic theories has significantly transformed our comprehension of the atomic and subatomic domains, revealing a complex array of phenomena regulated by the fundamental principles of quantum mechanics and QFT. The comprehensive framework for explaining the behavior of matter at its most fundamental levels has been facilitated by the concept of wave-particle duality, the significance of quantum numbers, and the arrangement of electron configurations.

As the exploration of the quantum universe progresses, contemporary atomic theories occupy a prominent position in scientific investigation, influencing our comprehension of the cosmos and serving as a source of inspiration for forthcoming generations of researchers seeking to unravel the quantum field's enigmas (Eckert, 2001).

Did You Know?

Ernest Rutherford's famous gold foil experiment in 1911 provided evidence for the existence of a dense, positively charged nucleus at the center of an atom, with electrons orbiting around it at a considerable distance.

2.12. SUMMARY

The chapter comprehensively explores the progressive development of our knowledge regarding the atom and its structural composition. The exploration commences with examining the extensive historical background of atomic theory, which can be traced back to the contributions of Democritus, an ancient Greek philosopher. Democritus introduced the notion of atoms, which are indivisible particles. Subsequently, an exploration ensues into the significant contributions made by John Dalton, who formulated the initial modern atomic theory during the early 19th century. Dalton's theory revolutionized the understanding of atoms by proposing their indivisible and identical nature, thereby establishing a fundamental framework for comprehending chemical combination principles and the concept of multiple proportions.

The subsequent chapter explores Niels Bohr's seminal atomic model, which was put forth in 1913. The model proposed by Bohr brought about a significant transformation in our comprehension of atomic behavior by introducing the notion of quantized energy levels and explaining the atomic spectrum of hydrogen. The postulates proposed by him regarding stationary orbits and quantized angular momentum served as a fundamental basis for the field of quantum mechanics, facilitating a connection between classical physics and quantum phenomena.

Isomerism is a phenomenon wherein compounds possessing identical chemical formulas manifest distinct structural configurations. The finding above has broadened our comprehension of molecular diversity and the complexities associated with chemical bonding. The chapter additionally explores the work of Dmitri Mendeleev, who contributed significantly to the field by developing the periodic table. This innovation involved the systematic arrangement of elements according to their properties and atomic masses. As a result, Mendeleev's periodic table enabled the anticipation of elements that had not yet been discovered and laid the foundation for establishing the periodic law. The phenomenon of isotopes was subsequently unveiled, explaining the presence of atoms belonging to the same chemical element but possessing varying neutron counts. The discovery presented in this study posed a challenge to the prevailing notion of uniform atomic masses and laid the groundwork for progress in nuclear chemistry.The subsequent chapter commences a comprehensive exploration of the identification and understanding of nuclear particles, encompassing electrons, protons, and neutrons. The cathode ray experiments conducted by J.J. Thomson were instrumental in the discovery and characterization of electrons. Similarly, Ernest Rutherford's gold foil experiment played a pivotal role in explaining the presence of the atomic nucleus. Lastly, James Chadwick's scientific investigations provided crucial evidence for the existence of neutrons, thereby contributing to a comprehensive understanding of atomic structure. Finally, the chapter explores the contemporary atomic theories, encompassing the notion of QFT and its utilization within the Standard Model of particle physics framework. These theories not only explain the behavior of atoms but also integrate fundamental forces and offer valuable insights into the subatomic domain. In summary, the chapter discussed the notable research of highly intelligent individuals in explaining the enigmas surrounding the atom and shaping our comprehension of the elemental particles of matter. By conducting an extensive examination of atomic models and theories, individuals are presented with diverse scientific concepts that persistently motivate and guide our ongoing pursuit to understand the complexities of the quantum field.

REFERENCES

1. Acevedo, J. F., Bramante, J., & Goodman, A., (2022). Accelerating composite dark matter discovery with nuclear recoils and the Migdal effect. *Physical Review D, 105*(2), 023012.
2. Amaral, L., Gopikrishnan, P., Matia, K., Plerou, V., & Stanley, H., (2001). Application of statistical physics methods and conceptsto the study of science & technology systems. *Scientometrics, 51*(1), 9–36.
3. Andriiko, A. A., & Lunk, H. J., (2018). The short form of Mendeleev's Periodic table of chemical elements: Toolbox for learning the basics of inorganic chemistry. A contribution to celebrate 150 years of the periodic table in 2019. *ChemTexts, 4*, 1–13.
4. Badash, L., (1983). Nuclear physics in Rutherford's laboratory before the discovery of the neutron. *American Journal of Physics, 51*(10), 884–889.
5. Baldes, R. W., (1978). 'Divisibility' and 'division' in Democritus. *Apeiron, 12*(1), 1–12.
6. Bor, N., (1985). The causality problem in atomic physics. *Uspekhi Fizicheskikh Nauk, 147*(2), 343–355.
7. Braune, S., (2010). From Lucretian atomic theory to Joycean Etymic theory. *Journal of Modern Literature, 33*(4), 167–181.
8. Büchler, H. P., Hermele, M., Huber, S. D., Fisher, M. P., & Zoller, P., (2005). Atomic quantum simulator for lattice gauge theories and ring exchange models. *Physical Review Letters, 95*(4), 040402.
9. Chalmers, A., (1998). Retracing the ancient steps to atomic theory. *Science & Education, 7*, 69–84.
10. Cheong, Y. W., & Song, J., (2014). Different levels of the meaning of wave-particle duality and a suspensive perspective on the interpretation of quantum theory. *Science & Education, 23*, 1011–1030.
11. Desrochers, J., Amrein, M. A., & Matyas, J. R., (2010). Structural and functional changes of the articular surface in a post-traumatic model of early osteoarthritis measured by atomic force microscopy. *Journal of Biomechanics, 43*(16), 3091–3098.
12. Dreyfuss, G., Swanson, M. S., & Piñol-Roma, S., (1988). Heterogeneous nuclear ribonucleoprotein particles and the pathway of mRNA formation. *Trends in Biochemical Sciences, 13*(3), 86–91.

13. Dunne, G. V., (2014). Extreme quantum field theory and particle physics with IZEST. *The European Physical Journal Special Topics*, *223*(6), 1055–1061.
14. Eaton, G. R., (2020). Whom should we credit for the discovery of isotopes? *Foundations of Chemistry*, *22*, 87–98.
15. Eckert, M., (2001). The emergence of quantum schools: Munich, Göttingen, and Copenhagen as new centers of atomic theory. *Annalen der Physik*, *513*(1–2), 151–162.
16. Eremin, I. E., Neshtimenko, V. V., Shcherban, D. S., & Fomin, D. V., (2020). Alternative relativistic formulation of linear spectrum of hydrogen-like systems. *Cybernetics and Physics*, *9*(4), 171–181.
17. Fafalis, D. A., Filopoulos, S. P., & Tsamasphyros, G. J., (2012). On the capability of generalized continuum theories to capture dispersion characteristics at the atomic scale. *European Journal of Mechanics-A/Solids*, *36*, 25–37.
18. Fakhri, H., & Hashemzadeh, R., (2009). Approach of the azimuthal and magnetic quantum numbers l and m to representation of a cubic deformation of Su(2) algebra. *International Journal of Modern Physics A*, *24*(25n26), 4727–4736.
19. Fano, U., (1956). Atomic theory of electromagnetic interactions in dense materials. *Physical Review*, *103*(5), 1202.
20. Feynman, R. P., (1954). Atomic theory of the two-fluid model of liquid helium. *Physical Review*, *94*(2), 262.
21. Furley, D., (1987). *The Greek Cosmologists: Volume 1, The Formation of the Atomic Theory and Its Earliest Critics* (Vol. 1, pp. 654–656). Cambridge University Press.
22. Ghosh, B., (2020). Scientific advancements leading to the evolution of the periodic table and discovery of new elements after Mendeleev. *Changing Trends in Human Thoughts and Perspectives: Science, Humanities, and Culture Part I*, *29*, 654–660.
23. Giunta, C. J., & Mainz, V. V., (2020). Discovery of nuclear magnetic resonance: Rabi, Purcell, and Bloch. In: *Pioneers of Magnetic Resonance* (Vol. 1, pp. 3–20). American Chemical Society.
24. Gol'Danskii, V. I., (1970). The periodic systems of DI Mendeleev and problems of nuclear chemistry. *Journal of Chemical Education*, *47*(6), 406. (translated by Avakian, P.).

25. Grossman, M. I., (2017). John Dalton and the origin of the atomic theory: Reassessing the influence of Bryan Higgins. *The British Journal for the History of Science*, *50*(4), 657–676.
26. Guo, H., Peng, Q., Chen, X. K., Gu, Q., Dong, S., Evans, E. W., & Li, F., (2019). High stability and luminescence efficiency in donor–acceptor neutral radicals not following the Aufbau principle. *Nature Materials*, *18*(9), 977–984.
27. Hartley, H. B., (1967). John Dalton, FRS (1766–1844) and the atomic theory – A lecture to commemorate his bicentenary. *Proceedings of the Royal Society of London. Series B. Biological Sciences*, *168*(1013), 335–359.
28. Hendry, J., (1980). The development of attitudes to the wave-particle duality of light and quantum theory, 1900–1920. *Annals of Science*, *37*(1), 59–79.
29. Hofer, W. A., Foster, A. S., & Shluger, A. L., (2003). Theories of scanning probe microscopes at the atomic scale. *Reviews of Modern Physics*, *75*(4), 1287.
30. Hon, G., & Goldstein, B. R., (2013). JJ Thomson's plum-pudding atomic model: The making of a scientific myth. *Annalen der Physik*, *525*(8–9), A129–A133.
31. Hunt, B. M., Li, J. I. A., Zibrov, A. A., Wang, L., Taniguchi, T., Watanabe, K., & Young, A. F., (2017). Direct measurement of discrete valley and orbital quantum numbers in bilayer graphene. *Nature Communications*, *8*(1), 948.
32. Jackman, T. E., Davies, J. A., Jackson, O. P., Unertl, W. N., & Norton, P. R., (1982). The Pt (110) phase transitions: A study by Rutherford backscattering, nuclear microanalysis, LEED, and thermal desorption spectroscopy. *Surface Science*, *120*(2), 389–412.
33. Karol, P. J., (2002). The Mendeleev-Seaborg periodic table: Through Z= 1138 and beyond. *Journal of Chemical Education*, *79*(1), 60.
34. Kedrov, B. M., (1949). Dalton's atomic theory and its philosophical significance. *Philosophy and Phenomenological Research*, *9*(4), 644–662.
35. Kibler, M. R., (2007). From the Mendeleev periodic table to particle physics and back to the periodic table. *Foundations of Chemistry*, *9*, 221–234.

36. Kragh, H., (1979). Niels Bohr's second atomic theory. *Historical Studies in the Physical Sciences*, *10*, 123–186.
37. Kuzmich, A., Bowen, W. P., Boozer, A. D., Boca, A., Chou, C. W., Duan, L. M., & Kimble, H. J., (2003). Generation of nonclassical photon pairs for scalable quantum communication with atomic ensembles. *Nature*, *423*(6941), 731–734.
38. Labuhn, H., Barredo, D., Ravets, S., De Léséleuc, S., Macrì, T., Lahaye, T., & Browaeys, A., (2016). Tunable two-dimensional arrays of single Rydberg atoms for realizing quantum Ising models. *Nature*, *534*(7609), 667–670.
39. Laing, M., (2008). The different periodic tables of Dmitrii Mendeleev. *Journal of Chemical Education*, *85*(1), 63.
40. Lal, D., (1988). In situ-produced cosmogenic isotopes in terrestrial rocks. *Annual Review of Earth and Planetary Sciences*, *16*(1), 355–388.
41. Leone, M., Robotti, N., & Verna, G., (2018). 'Rutherford's experiment' on alpha particles scattering: The experiment that never was. *Physics Education*, *53*(3), 035003.
42. Lovisetti, L., & Giliberti, M., (2022). "I (critically) think, therefore I am": Thomson's atomic model and the ineffectiveness of physics education. In: *Proceedings of the IUPAP International Conference on Physics Education* (Vol. 2, pp. 116–116).
43. Makhmudov, A. A., & Khudayberganov, A. M., (2023). Methods of teaching the topic "Bohr's postulates" of the section "Atomic physics" of the course of general physics in higher educational institutions. *European Journal of Innovation in Nonformal Education*, *3*(2), 1–8.
44. Mátyus, E., & Reiher, M., (2012). Molecular structure calculations: A unified quantum mechanical description of electrons and nuclei using explicitly correlated Gaussian functions and the global vector representation. *The Journal of Chemical Physics*, *137*(2), 6510.
45. Metz, I. K., Bennett, J. W., & Mason, S. E., (2021). Examining the Aufbau principle and ionization energies: A computational chemistry exercise for the introductory level. *Journal of Chemical Education*, *98*(12), 4017–4025.
46. Mills, R. L., (2001). The nature of free electrons in superfluid helium – a test of quantum mechanics and a basis to review its foundations and make a comparison to classical theory. *International Journal of Hydrogen Energy*, *26*(10), 1059–1096.

47. Mirny, L., & Shakhnovich, E., (2001). Protein folding theory: From lattice to all-atom models. *Annual Review of Biophysics and Biomolecular Structure*, *30*(1), 361–396.
48. Moore, E. A., (1999). Relativistic chemical shielding: formally exact solutions for one—electron atoms of maximum total angular momentum for any principal quantum number. *Molecular Physics*, *97*(3), 375–380.
49. Murray, J., (1819). XVII. Observations on the relation of the law of definite proportions in chemical combination, to the constitution of the acids, alkalis, and earths. *The Philosophical Magazine*, *54*(256), 90–100.
50. Niaz, M., (2001). How important are the laws of definite and multiple proportions in chemistry and teaching chemistry? A history and philosophy of science perspective. *Science & Education*, *10*, 243–266.
51. Oganessian, Y., (2019). Super heavy elements: On the 150th anniversary of the discovery of the periodic table of elements. *Nuclear Physics News*, *29*(1), 5–10.
52. Peterman, K., & Cordes, M., (2021). England: John Dalton and the atomic theory. In: *Contextualizing Climate Change: Linking Science and Culture* (Vol. 2, pp. 37–47). American Chemical Society.
53. Pratt, H. T., (2010). A letter signed: the very beginnings of Dalton's atomic theory. *Ambix*, *57*(3), 301–310.
54. Primas, H., (1975). Pattern recognition in molecular quantum mechanics: I. Background dependence of molecular states. *Theoretica Chimica Acta*, *39*, 127–148.
55. Rico, E., Pichler, T., Dalmonte, M., Zoller, P., & Montangero, S., (2014). Tensor networks for lattice gauge theories and atomic quantum simulation. *Physical Review Letters*, *112*(20), 201601.
56. Rocke, A. J., (2005). In search of El Dorado: John Dalton and the origins of the atomic theory. *Social Research: An International Quarterly*, *72*(1), 125–158.
57. Roller, D. H., (1981). Greek atomic theory. *American Journal of Physics*, *49*(3), 206–210.
58. Sabbatucci, L., & Salvat, F., (2016). Theory and calculation of the atomic photoeffect. *Radiation Physics and Chemistry*, *121*, 122–140.
59. Sansone, A., Melchiorre, M., Chatgilialoglu, C., & Ferreri, C., (2013). Hexadecenoic fatty acid isomers: A chemical biology approach

for human plasma biomarker development. *Chemical Research in Toxicology*, *26*(11), 1703–1709.

60. Sarathy, S. M., Vranckx, S., Yasunaga, K., Mehl, M., Oßwald, P., Metcalfe, W. K., & Curran, H. J., (2012). A comprehensive chemical kinetic combustion model for the four butanol isomers. *Combustion and Flame*, *159*(6), 2028–2055.
61. Schwerdtfeger, P., Smits, O. R., & Pyykkö, P., (2020). The periodic table and the physics that drives it. *Nature Reviews Chemistry*, *4*(7), 359–380.
62. Selco, J., Bruno, M., & Chan, S., (2013). Discovering periodicity: Hands-on, minds-on organization of the Periodic Table by visualizing the unseen. *Journal of Chemical Education*, *90*(8), 995–1002.
63. Shaik, S., (2007). The Lewis legacy: The chemical bond – A territory and heartland of chemistry. *Journal of Computational Chemistry*, *28*(1), 51–61.
64. Shao, Y., Molnar, L. F., Jung, Y., Kussmann, J., Ochsenfeld, C., Brown, S. T., & Head-Gordon, M., (2006). Advances in methods and algorithms in a modern quantum chemistry program package. *Physical Chemistry Chemical Physics*, *8*(27), 3172–3191.
65. Sigmund, P., & Haagerup, U., (1986). Bethe stopping theory for a harmonic oscillator and Bohr's oscillator model of atomic stopping. *Physical Review A*, *34*(2), 892.
66. Steffens, A., Friesdorf, M., Langen, T., Rauer, B., Schweigler, T., Hübener, R., & Eisert, J., (2015). Towards experimental quantum-field tomography with ultracold atoms. *Nature Communications*, *6*(1), 7663.
67. Sylvester, J. J., (1878). On an application of the new atomic theory to the graphical representation of the invariants and covariants of binary quantics, with three appendices. *American Journal of Mathematics*, *1*(1), 64–104.
68. Thackray, A. W., (1966). The origin of Dalton's chemical atomic theory: Daltonian doubts resolved. *Isis*, *57*(1), 35–55.
69. Tosoni, S., Di Liberto, G., Matanovic, I., & Pacchioni, G., (2023). Modelling single atom catalysts for water splitting and fuel cells: A tutorial review. *Journal of Power Sources*, *556*, 232492.
70. Turner, P. S., & Rowe, D. J., (2005). Phase transitions and quasidynamical symmetry in nuclear collective models. II. The spherical vibrator to gamma-soft rotor transition in an SO(5)-invariant Bohr model. *Nuclear Physics A*, *756*(3–4), 333–355.

71. Van Tiggelen, B., & Lykknes, A., (2019). Celebrate the women behind the periodic table. *Nature*, *565*(7741), 559–561.
72. Wallace, D., (2011). Taking particle physics seriously: A critique of the algebraic approach to quantum field theory. *Studies in History and Philosophy of Science Part B: Studies in History and Philosophy of Modern Physics*, *42*(2), 116–125.
73. Westermayr, J., & Marquetand, P., (2020). Deep learning for UV absorption spectra with SchNarc: First steps toward transferability in chemical compound space. *The Journal of Chemical Physics*, *153*(15), 651–655.
74. Willett, R., Eisenstein, J. P., Störmer, H. L., Tsui, D. C., Gossard, A. C., & English, J. H., (1987). Observation of an even-denominator quantum number in the fractional quantum Hall effect. *Physical Review Letters*, *59*(15), 1776.

CHAPTER 3

Atomic Spectra and Emissions

Contents

3.1. INTRODUCTION

The study of spectra and emissions can be traced back to the latter part of the 19th century and early 20th century. During this period, physicists achieved notable advancements in comprehending the atomic structure and the behavior of atoms under diverse circumstances. Prominent figures who have significantly contributed to this discipline include Max Planck, Niels Bohr, and Werner Heisenberg. Their seminal research has established the fundamental principles upon which the advancement of quantum mechanics and its subsequent utilization in atomic phenomena have been built (Parigger, 2013).

Atomic spectroscopy fundamentally encompasses examining the interactions between atoms and electromagnetic radiation, encompassing phenomena such as the absorption, emission, and scattering of photons. Investigating atomic spectra and emissions is important across various scientific fields, encompassing astrophysics, chemistry, materials science, and medical diagnostics. The advancement of scientific knowledge has facilitated the exploration and understanding of various domains, including identifying elements, chemical analysis, investigating celestial bodies, and advancing sophisticated technologies (Savall-Alemany et al., 2016).

To understand the complex behavior exhibited by atoms, it is imperative to explore the domain of quantum mechanics. The inadequacy of classical physics in accounting for various phenomena observed in atomic systems has prompted the emergence of quantum mechanics as a novel theoretical framework. The discipline of physics has contributed to the development of the concept of quantization, which entails the confinement of specific attributes of particles, such as energy levels, to discrete values instead of continuous ranges.

The role of energy levels in atomic spectroscopy is of fundamental significance. When an atom engages in an interaction with electromagnetic radiation, the electrons within the atom can absorb energy and undergo a transition to energy levels that are of a higher magnitude.

Consequently, when these electrons transition to lower energy levels, they release photons with distinct wavelengths, creating atomic spectra. The spectra exhibit discernible lines corresponding to the energies of the emitted photons, thereby functioning as unique identifiers for individual atoms. Consequently, these spectral patterns facilitate the identification and investigation of atoms (Aguilera & Aragón, 2004). There are two main categories of atomic spectra: continuous and line spectra. A continuous spectrum is observed when a wide range of wavelengths is emitted or absorbed, resulting in a seamless continuum of colors. On the other hand, a line spectrum is characterized by discrete and precisely defined spectral lines, with each line corresponding to a distinct energy level transition within an atom. The unique characteristics exhibited by line spectra render them highly valuable in the field of elemental analysis and in the identification of specific elements within samples.

In terms of energy absorption and release, the atomic spectra can be observed in two distinct forms: the absorption spectrum and the emission spectrum. An absorption spectrum occurs when an atom selectively absorbs distinct wavelengths of light, resulting in transitions from lower to higher energy states. Consequently, absorption lines, characterized by dark lines, are observed in the spectrum. Conversely, the emission spectrum arises from the phenomenon wherein electrons that have been excited within an atom subsequently transition to lower energy states, thereby releasing photons with distinct energies. The emitted photons give rise to luminous lines, commonly referred to as emission lines, within the spectrum.

The mechanisms that govern the emission spectra in atoms are the fundamental basis for numerous technological advancements. An example of a notable occurrence is the stimulated emission of radiation, which serves as the fundamental principle behind the operation of lasers. Stimulated emission is a phenomenon where the emission of photons is induced by an incoming photon, resulting in the release of additional photons by pre-existing excited atoms. This process leads to the generation of coherent and amplified light output. The procedure above holds significant importance in many applications, encompassing laser technology, optical communications, and medical devices (Yngström, 1994).

The Bohr model attempted to investigate the atomic spectra and emissions. The proposal made by Niels Bohr posits that electrons within atoms are situated at discrete energy levels, and the transition between these levels leads to the emission or absorption of photons possessing distinct energies. Although the Bohr model effectively discussed the hydrogen spectrum, it exhibited certain constraints and proved inadequate in comprehensively characterizing the behavior of atoms with greater complexity.

The field of quantum mechanics has significantly transformed the comprehension of atomic spectra by providing a more comprehensive and precise depiction of atomic behavior. The Schrödinger equation, a pivotal equation in quantum mechanics, facilitates the computation of wavefunctions that characterize the likelihood distribution of electrons in the vicinity of an atomic nucleus. The mathematical formalism presented herein offers a rigorous explanation of atomic spectra, with quantum numbers governing atomic orbitals' configuration and energy states.

The introduction of quantum mechanics expanded the comprehension of atomic spectra and emissions to encompass elements other than hydrogen, thereby facilitating a deeper understanding of the periodic table and the behavior of diverse elements in varying environments. The utilization of quantum mechanical models has emerged as a fundamental aspect of contemporary atomic spectroscopy, facilitating the examination and anticipation of intricate atomic and ionic spectra under various circumstances (Parsons et al., 1966).

Atomic spectroscopy utilizes a range of methodologies to investigate atomic spectra and emissions in practical applications. Optical spectroscopy, encompassing absorption and emission spectroscopy, is a prevalent technique to analyze the interplay between atoms and light. Absorption spectroscopy involves quantifying light absorption by an atom or sample across various wavelengths, thereby explaining absorption lines that correspond to energy level transitions of the atoms. On the other hand, emission spectroscopy is a technique that quantifies the intensity of emitted light about its wavelength, thereby generating emission lines that yield significant insights into the energy levels of the atom.

Mass spectrometry is a highly significant spectroscopic technique that enables the determination of an atom's mass and elemental composition by analyzing the deflection of ions in the presence of a magnetic field. Mass spectrometry is utilized in various fields, such as isotopic analysis, identifying pollutants in environmental samples, and dating archaeological artifacts (Hermsmeier et al., 1988).

Tip

Atomic emission spectra are like unique "fingerprints" for elements, allowing scientists to identify elements in distant stars based on the light they emit.

3.2. ATOMIC SPECTRA

The arrangement of electrons within an atom is typically such that the overall energy of the atom is minimized. The ground state of an atom refers to the state with the lowest energy level. When provided with energy, the atoms undergo a process wherein the electrons assimilate the energy and transition to a higher energy state.

The energy levels of electrons in atoms exhibit quantization, indicating that electron transitions between energy levels occur in discrete increments rather than continuously. An atom in an excited state refers to a condition in which its potential energy exceeds the ground states. An atom in an excited state exhibits a lack of stability. Upon transitioning back to its ground state, it expels the energy it had previously acquired in the form of electromagnetic radiation (Layzer, 1959).

The application of an electric current to a confined gas sample at a reduced pressure exhibits a distinct behavior. Due to the distinct electron energy levels associated with each element, the emission of light in a gas discharge tube is characterized by a discernible color depending upon the specific gas present (Figure 3.1).

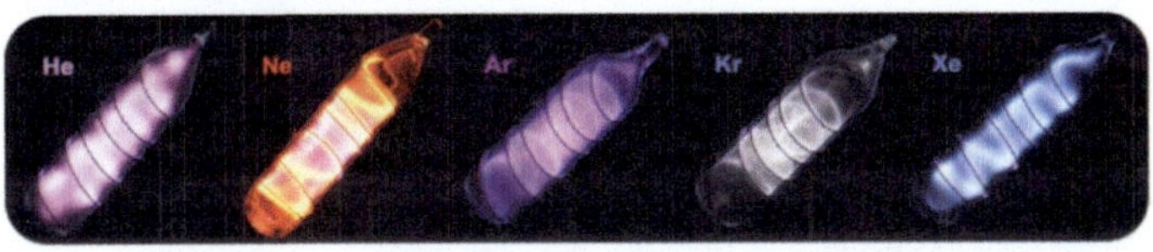

Figure 3.1. *Gas discharge tubes are enclosed glass tubes filled with a gas at low pressure, through which an electric current is passed.*

Source: Heinrich, Creative Commons License.

Gas discharge tubes, such as "neon" signs, are commonly recognized illustrations. Nevertheless, it is important to note that only signs exhibiting the red-orange hue depicted in the figure are infused with neon. Signs of varying colors are indicative of distinct gas compositions or gas mixtures.

The researchers investigated the unique pink hue exhibited by the gas discharge generated by hydrogen gas. Upon subjecting a concentrated beam of this luminous radiation to the process of refraction through a prism, the resultant outcome was the division of the light into four distinct lines characterized by highly precise wavelengths (and correspondingly, frequencies, as the variables λ and ν exhibit an inverse relationship). The atomic emission spectrum refers to the characteristic arrangement of spectral lines that arise when light undergoes dispersion through a prism, separating its constituent wavelengths or frequencies. The diagram presented in Figure 3.2 illustrates the atomic emission spectrum of hydrogen (Sobel'Man, 2016).

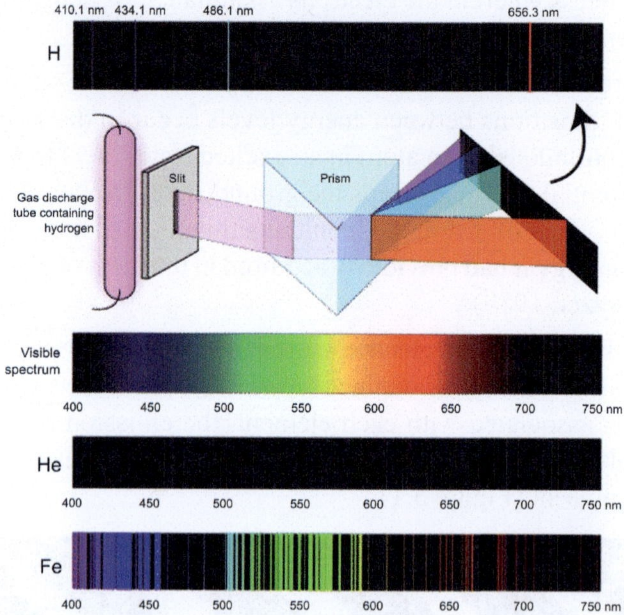

Figure 3.2. *When light from a hydrogen discharge tube is passed through a prism, the light splits into four visible lines. Each of these spectral lines corresponds to a different electron transition from a higher energy state to a lower energy state.*

Source: Heinrich, Creative Commons License.

The classical theory did not explain the presence of atomic emission spectra, commonly called line-emission spectra. Based on the principles of classical physics, it is postulated that an atom in its ground state can absorb energy continuously, as opposed to being restricted to discrete energy levels. Similarly, when the atoms returned to a state of lower energy, they had the potential to release varying amounts of energy. This phenomenon would give rise to a continuous spectrum encompassing the representation of all wavelengths and frequencies. The atomic emission spectra provided further evidence for the discrete nature of light and subsequently contributed to developing a new atomic model grounded in the principles of quantum theory.

The investigation of the interaction between atoms and electromagnetic radiation, known as atomic spectra, has been a fundamental aspect of contemporary physics since the inception of the 20th century. Observing discrete spectral lines emitted or absorbed by atoms when exposed to light or other electromagnetic radiation has significantly transformed our comprehension of atomic structure and the underlying principles of quantum mechanics.

The investigation of atomic spectra can be traced back to the 19th century, during which scientists initiated the examination of light behavior when subjected to prisms and diffraction gratings. Researchers were intrigued by continuous spectra, exemplified by the rainbow of colors that emerges when sunlight is refracted through a prism. Nevertheless, it was the analysis of emission spectra that led to a significant revelation (Wigner, 2012).

During the mid-1800s, German physicists Gustav Kirchhoff and Robert Bunsen conducted a series of groundbreaking experiments. These experiments focused on analyzing spectral lines emitted by various chemical elements when subjected to heat. It was determined that each element displayed a distinct arrangement of luminous lines at precise wavelengths. The finding above established the fundamental basis for comprehending atomic spectra and the notion of atomic fingerprints, wherein each element possesses a unique collection of spectral lines.

The pursuit of explaining the genesis of atomic spectra resulted in a seminal advancement in contemporary physics, namely the formulation of the Bohr model of the atom. The Bohr model, first proposed by Niels Bohr in 1913, effectively accounted for the observed spectral lines of hydrogen and served as a foundational milestone in developing quantum mechanics (Cowan, 1968).

According to the Bohr model, electrons within an atom are confined to specific energy levels or orbits encircling the nucleus. When an electron transitions from a state of higher energy to a state of lower energy, it releases a photon with a distinct energy value corresponding to the difference in energy between the two states above. The photons emitted are responsible for the distinct spectral lines observed in atomic emission spectra.

Despite its revolutionary nature, the Bohr model was subject to limitations that hindered its ability to account for intricate features in spectral lines or the phenomenon of line splitting observed under specific circumstances. Consequently, it established the groundwork for advancing a more extensive quantum mechanical representation of the atom (Kaufman & Edlén, 1974).

The field of quantum mechanics, which emerged during the 1920s, offered a deeper comprehension of atomic phenomena and spectral characteristics. Fundamentally, quantum mechanics discusses the dynamics of matter and energy at the microscopic level, wherein classical physics proves inadequate in offering a precise depiction.

In the context of the quantum mechanical model of the atom, it is observed that electrons do not exhibit confinement to discrete orbits. Instead, their behavior is described by wavefunctions, which serve to characterize the likelihood of locating an electron within a specific region surrounding the atomic nucleus. The wavefunctions presented herein indicate discrete energy levels, giving rise to the quantized characteristics observed in atomic spectra (Rosenzweig & Porter, 1960).

Energy level diagrams illustrate the permissible transitions between different energy levels, while selection rules dictate which transitions are permissible. The rules above govern the alterations in quantum numbers that occur during electron transitions and establish the observable spectral lines within atomic spectra.

Spectroscopy is an academic discipline that examines the intricate relationship between matter and electromagnetic radiation. It is a highly effective instrument for exploring and analyzing atomic spectra. The process entails the examination of light that is either transmitted or emitted to derive significant insights regarding the characteristics of atoms and molecules (Suenaga & Koshino, 2010).

Absorption spectroscopy is a technique used to quantify the extent of light absorption exhibited by a given sample across a range of wavelengths. The sample is exposed to a continuous spectrum of light, and certain wavelengths that align with the energy level transitions of the atoms are preferentially absorbed. Consequently, the presence of dark absorption lines becomes evident in the spectrum, thereby offering valuable insights into the energy states of the atoms (Kalfas et al., 2016).

Remember

The spectral lines observed in atomic emission spectra have been used to develop fun, interactive tools like spectroscope toys that let you see the colorful patterns of light from everyday objects.

3.3. TYPES OF ATOMIC SPECTRA

The field of atomic spectra investigates the complex dynamics that arise from the interaction between atoms and electromagnetic radiation, shedding light on the intricate nature of the quantum field. When atoms interact with light or other forms of electromagnetic radiation, they generate unique spectral patterns, which provide valuable insights into their energy levels and electron transitions. The atomic spectra can be classified into three primary categories: continuous, emission, and absorption spectra. These distinct types of spectra provide valuable insights into the behavior of atoms and the fundamental principles of quantum mechanics. This analysis aims to thoroughly examine various types of atomic spectra, explaining their distinctive attributes and their profound implications in comprehending atomic behavior (Hoenig & de Kersabiec, 1996) (Figure 3.3).

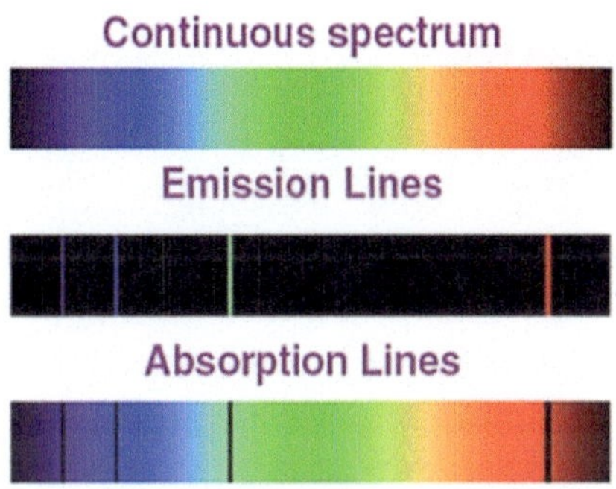

Figure 3.3. *Illustration of atomic spectra.*

Source: BYJU'S, Creative Commons License.

3.3.1. Continuous Spectra

Continuous spectra, also known as a continuous spectrum, are observed when a broad range of wavelengths or frequencies of electromagnetic radiation is emitted or absorbed by an atom. These spectra typically result from the thermal motion of particles or the interaction of light with a dense, hot material (Hong et al., 2017). An excellent example of a continuous spectrum is the emission of visible light from a hot, solid object, such as a metal rod, when it is heated. In this case, a seamless rainbow of colors is produced, with no distinct lines or gaps representing a continuous distribution of wavelengths (Garcia & Mack, 1965).

3.3.2. Emission Spectra

On the other hand, line spectra, also known as discrete spectra or emission spectra, exhibit distinct and well-defined lines or discrete wavelengths at which an atom emits or absorbs light. These lines correspond to specific energy transitions within the atom, typically involving electrons moving between energy levels or orbitals. Line spectra are a hallmark of individual elements and provide a unique "fingerprint" for each element, making them essential tools in chemical analysis and the identification of substances. The most famous example is the hydrogen emission spectrum, which consists of a series of bright lines in the visible spectrum, each corresponding to a specific electron transition within the hydrogen atom (Bauche-Arnoult et al., 1982; Chinni, 2012).

3.3.3. Absorption Spectrum

The absorption spectrum can be considered reciprocal to the emission spectrum. Atoms absorb particular wavelengths of light during electron transitions from lower to higher energy levels rather than emitting light. When a sample of atoms is exposed to a continuous spectrum of light, specific wavelengths are absorbed by the atoms in a selective manner, manifesting dark lines against a bright background.

Dark absorption lines in the spectrum can be attributed to the absorption of photons by atoms, resulting in the excitation of electrons to higher energy levels. Absorption spectra are commonly acquired through light transmission across a specimen composed of gas, liquid, or solid substances. When light interacts with atoms, certain wavelengths are selectively absorbed, forming distinct dark lines that possess unique characteristics associated with the absorbing atoms (Lewen, 2011). The Bohr model additionally offered a theoretical framework for explaining the hydrogen absorption spectrum phenomenon. When an atom absorbs a photon with energy equivalent to the difference in energy levels, an electron transitions from a lower to a higher energy level. The observed dark absorption lines in the spectrum can be attributed to the transitions between different energy levels (Solovyev et al., 2020).

3.3.4. Hydrogen Atom and Balmer Series

Due to its comparatively uncomplicated electronic configuration, the hydrogen atom has been of significant importance in examining atomic spectra. The Balmer series is a widely recognized illustration of the spectral lines discharged by hydrogen. The Balmer series encompasses a collection of distinct emission lines that are prominently visible within the electromagnetic spectrum.

The Balmer series is a spectral series that originates from electron transitions within hydrogen atoms, specifically between energy levels that are greater than 2 ($n > 2$) or the second energy level ($n = 2$). When an electron transitions from a higher energy level to the second energy level, it releases a photon possessing a distinct energy that corresponds to the disparity in energy between the two levels. The emitted photons manifest themselves as luminous lines within the Balmer series.

The Balmer series holds notable importance because its spectral lines are observable by the human eye. The emission of photons within the visible portion of the electromagnetic spectrum occurs when electrons transition to the second energy level from energy levels of higher magnitude, leading to the formation of distinct and easily detectable spectral lines. The Balmer series constituted an early body of evidence that lent support to the Bohr model of the atom, thereby reinforcing the significance of atomic spectra in comprehending atomic behavior (Onorato et al., 2015).

3.4. QUANTUM MECHANICS AND THE FINE STRUCTURE

Although the Bohr model made significant strides in explaining the emission spectrum of the hydrogen atom, its applicability to more intricate atoms was constrained. As the investigation into atomic spectra advanced, it became apparent that a more encompassing model was imperative to discuss the intricate nuances observed in spectral lines. The field of quantum mechanics, which emerged during the 1920s, brought about a significant paradigm shift in the comprehension of atomic phenomena and the characteristics of spectral patterns. The concept of wave-particle duality was introduced, wherein particles such as electrons can manifest characteristics of both waves and particles. In quantum mechanics, the behavior of electrons deviates from the notion of being restricted to definite orbits. Instead, their characteristics are discussed through wavefunctions, which depict the likelihood of locating an electron within a specific vicinity surrounding the nucleus (Askerka et al., 2017). The utilization of quantum mechanics in the analysis of the hydrogen atom yielded a more precise depiction of the emission spectrum, encompassing the phenomenon of fine structure splitting. The term "fine structure" pertains to the minute disparities in energy levels of electrons that the Bohr model does not adequately explain. The field of quantum mechanics has provided insights into the behavior of electrons within an atom, demonstrating that they undergo supplementary interactions, such as the spin-orbit interaction. As a consequence of these interactions, the spectral lines emitted by atoms are observed to split into multiple components (Marsman et al., 2015).

3.5. THE ZEEMAN EFFECT

The Zeeman effect, first observed by Pieter Zeeman, a prominent Dutch physicist, in 1896, is an intriguing phenomenon within atomic spectra. The Zeeman effect pertains to the division of spectral lines when a magnetic field is present. When atoms are exposed to a magnetic field, the magnetic interaction between them and their electrons leads to the division of energy levels, thereby giving rise to multiple constituents of a solitary spectral line (Figure 3.4).

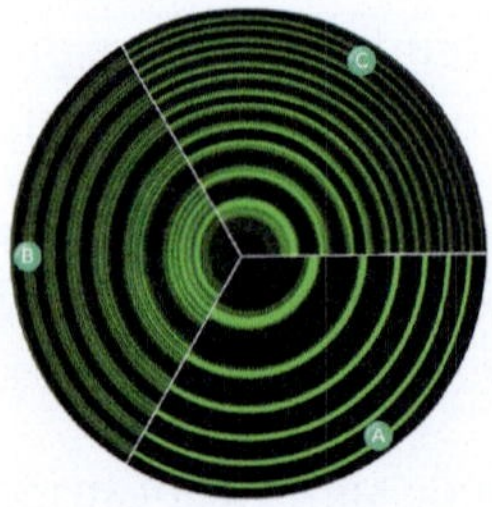

Figure 3.4. *Representation of the Zeeman effect.*

Source: McCarthy, Creative Commons License.

The Zeeman effect arises due to the interplay between the magnetic moments of electrons and an external magnetic field. Under the influence of a week magnetic field, referred to as the weak-field Zeeman effect, the spectral lines undergo a division into three distinct components: the central component, which corresponds to the original wavelength, and two side components that exhibit a shift towards longer and shorter wavelengths, respectively. In the context of the strong-field Zeeman effect, it is observed that the spectral lines exhibit a phenomenon wherein they undergo a splitting into numerous components, resulting in the emergence of an intricate pattern consisting of multiple lines. The Zeeman effect is highly effective for examining atomic characteristics, including electron spin and magnetic moments. This phenomenon finds practical applications in various fields, such as spectroscopy, astrophysics, and plasma diagnostics (Berdyugina & Solanki, 2002).

3.6. THE STARK EFFECT

Like the Zeeman effect, the Stark effect is a fundamental phenomenon observed in atomic spectra. The Stark effect pertains to spectral line splitting and shifting occurring when an electric field is present. Applying an electric field to atoms results in the disturbance of electron energy levels, which subsequently causes alterations in the positions and intensities of spectral lines (Figure 3.5).

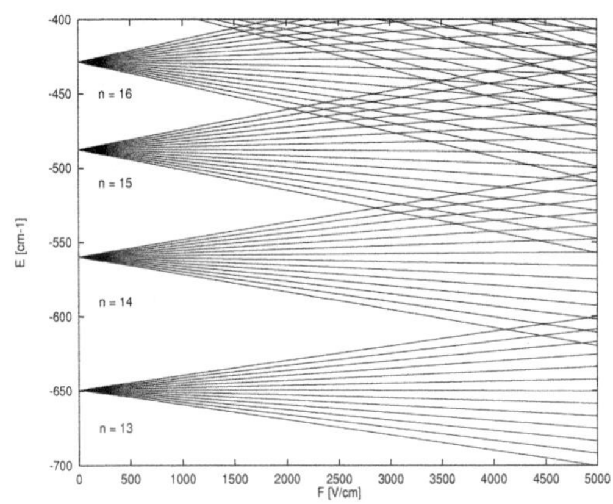

Figure 3.5. *Graphical representation of Stark effect.*

Source: Wikipedia, Creative Commons License.

The Stark effect can be attributed to the interaction between an external electric field and the electric dipole moment of an atom. This interaction arises due to the displacement of the electron cloud about the nucleus. Like the Zeeman effect, the Stark effect can induce intricate splitting and displacement of spectral lines, offering significant insights into atomic structure and interactions (Epstein, 1926).

The study of atomic spectra has provided valuable insights into the quantum field by revealing the intricate behavior of electrons within atoms. The investigation of atomic behavior in modern quantum mechanics is based on analyzing three primary types of atomic spectra: continuous, emission, and absorption. These spectra provide valuable insights into different aspects of atomic behavior.

The presence of seamless and unbroken color gradients characterizes continuous spectra. These spectra originate from an infinite range of wavelengths that are emitted or absorbed by atoms within a highly dense and heated medium. Emission spectra are distinguished by a series of distinct lines, which arise from releasing photons during the transition of electrons from higher to lower energy states. Absorption spectra are characterized by dark lines against a bright background, which arises due to the absorption of particular wavelengths of light by atoms during the transition of electrons from lower to higher energy levels (Fried & Boxer, 2015).

Investigating atomic spectra has resulted in the formulation of quantum mechanics. This fundamental theoretical framework discusses the behavior of matter and energy at the atomic and subatomic scales. The Bohr model of the atom initially contributed to the comprehension of atomic emission spectra. In contrast, quantum mechanics subsequently discussed the phenomenon of fine structure splitting and introduced the concept of wave-particle duality, thereby unveiling the intricate characteristics of electrons within atomic systems.

The Zeeman and Stark effects are observed in atomic spectra when subjected to magnetic and electric fields, respectively. These effects have significantly contributed to advancing the knowledge regarding atomic properties and interactions (Hehenberger et al., 1974).

Did You Know?

Atomic emission spectroscopy can detect trace elements in water as low as parts per billion, making it crucial for environmental monitoring and quality control.

3.7. EMISSION PROCESSES IN ATOMS

The emission processes observed in atoms entail the emission of electromagnetic radiation, specifically photons, due to electron transitions from higher to lower energy levels. These phenomena are of great significance in atomic and quantum physics, as they offer an essential understanding of the energy arrangement of atoms and the fundamental principles of quantum mechanics. A comprehensive overview of emission processes is important across various fields, encompassing astrophysics, spectroscopy, and the advancement of lasers and optoelectronic devices (Kulander & Rescigno, 1991).

3.7.1. Electron Energy Levels and Transitions

The fundamental basis of emission processes is rooted in electron energy levels within atoms. During the initial decades of the 20th century, Niels Bohr introduced his atomic model, which proposed that electrons within an atom are situated in discrete energy levels or orbits encircling the nucleus. Based on Bohr's model, electrons closer to the atomic nucleus are found to occupy energy levels of lower magnitude. In contrast, electrons situated at greater distances from the nucleus occupy energy levels of higher magnitude.

In the ground state of an atom, all of its electrons occupy the energy levels with the lowest possible energy. Nevertheless, if the atom is subjected to external energy, such as heat or light, a portion of its electrons can assimilate this energy and transition to energy levels of greater magnitude. The state above is commonly referred to as an excited state, which typically exhibits a transient nature due to the inherent instability of electrons at higher energy levels (Leapman et al., 1982).

3.7.2. Emission of Photons

When electrons in an excited state transition back to lower energy levels, they emit surplus energy in the form of photons, discrete units of electromagnetic radiation. Every individual photon that is released possesses an amount of energy corresponding to the disparity in energy between the two distinct levels of energy engaged in the transition. This is commonly referred to as emission. The emission of photons is regulated by the principles of quantum mechanics, wherein energy levels are quantized, implying that they can solely assume distinct, discrete values. Energy quantization holds significant importance in atomic physics, as it has profound implications for the behavior of atoms and matter at both the atomic and subatomic levels (Nilsson & Kröll, 2005).

3.7.3. Spectroscopic Notation

The field of spectroscopy employs a notation system to discuss the energy states of atoms and the interchanges occurring between these states. The notation system utilizes a combination of alphanumeric characters to symbolize the principal quantum number (n) and other quantum numbers linked to each energy level. The principal quantum number (n) denotes the primary energy level or orbital in which the electron is located. This variable possesses a discrete range of values commencing from the integer 1, wherein larger values correspond to greater energy levels. Additional quantum numbers, namely the azimuthal (l) and the magnetic quantum number (ml), delineate the distinct sublevels within each energy level (Schutte et al., 1997).

3.7.4. Permitted and Prohibited Transitions

In the emission process, it is important to note that not all transitions between energy levels are permissible. The field of quantum mechanics enforces selection rules that dictate the permissible transitions, delineating the probable and prohibited transitions. The predominant criterion for the emission process is that the alteration in the principal quantum number (Δn) must exhibit a value of +1. Stated differently, the transition of electrons is limited to descending from energy levels of higher quantum numbers (n+1) to those of lower quantum numbers (n). The principle above is commonly called the "one-photon rule" due to its association with the emission of a solitary photon that carries the energy difference between the above-mentioned levels (Fujisawa et al., 2002).

3.7.5. Spin Selection Rule

The spin selection rule is an additional criterion that regulates emission processes. Electrons possess an inherent characteristic known as spin, which exhibits a binary nature with two potential states: +½ or –½. Based on the spin selection rule, the electron's spin must remain unaltered throughout the emission process.

This implies that electrons possessing a spin of +½ are exclusively capable of transitioning to states characterized by a spin of +½. In contrast, electrons with a spin of –½ can solely transition to states characterized by a spin of –½ (Ezawa, 2012).

3.7.6. Emission Lines and Spectra

Distinct lines in an emission spectrum are generated due to the emission of photons during transitions between energy levels. Each line corresponds to a distinct transition between energy levels, and the energy of the emitted photons determines the wavelength or frequency of the spectral line. Hydrogen, a highly investigated element in atomic physics, exhibits the Balmer series as a prominent illustration of an emission spectrum.

The Balmer series encompasses a collection of distinct spectral lines prominently observed within the visible region of the electromagnetic spectrum. These spectral lines indicate electron transitions occurring between energy levels of greater magnitude and the second energy level of the atom (Kennicutt Jr, 1992).

3.7.7. Spontaneous Emission

The phenomenon known as spontaneous emission refers to the emission of photons by atoms in an excited state without any external influence or stimulation. When an atom is in an excited state, it possesses inherent instability, leading to the eventual spontaneous return of its electrons to lower energy levels, accompanied by the emission of photons.

Spontaneous emission is an intrinsic phenomenon that manifests in diverse contexts, including the emission of light by atoms and molecules in excited states, the emission of photons by celestial bodies, and the decay of radioactive isotopes. The phenomenon above holds considerable importance in comprehending atomic behavior and radiative mechanisms within the cosmos (Noda et al., 2007).

Stimulated emission is a significant phenomenon observed in atomic systems, pivotal in the functioning of laser devices. In contrast to spontaneous emission, stimulated emission occurs when external electromagnetic radiation is present.

Stimulated emission refers to the phenomenon wherein an incident photon interacts with an excited atom, resulting in the atom transitioning from a higher to a lower energy level. As a consequence of this transition, a second photon is emitted, which possesses identical characteristics to the incident photon in terms of energy, phase, and direction of propagation. This procedure yields light intensification, generating coherent and highly concentrated radiation beams.

The fundamental basis of laser technology lies in the capacity to produce and manipulate stimulated emission. Lasers, an abbreviation for "Light Amplification by Stimulated Emission of Radiation," operate based on stimulated emission principles to generate concentrated and potent light beams. Stimulated emission is pivotal in various applications, encompassing telecommunications, materials processing, medical procedures, and research endeavors (Friedman et al., 1988).

The emission processes occurring in atoms entail the emission of electromagnetic radiation, specifically photons, as a consequence of electron transitions from higher to lower energy levels. The processes above are of utmost importance in atomic and quantum physics, as they offer invaluable perspectives on the energy configuration of atoms and the fundamental principles governing quantum mechanics. The allowed transitions and spectral lines observed in emission spectra are governed by spectroscopic notation and selection rules. The Balmer series about hydrogen is widely recognized as an illustrative instance of an emission spectrum. Spontaneous emission is a phenomenon that arises when atoms in an excited state transition to lower energy levels without any external influence. On the other hand, stimulated emission, which plays a crucial role in the operation of lasers, involves the amplification of light resulting from the interaction between incoming photons and atoms in an excited state. The comprehension of emission processes facilitates the explanation of the intricate movement of electrons within atoms, thereby enhancing the understanding of atomic behavior and the fundamental principles underlying the quantum field (Hell & Wichmann, 1994).

Remember

The energy level diagrams used to depict atomic transitions can be thought of as "ladders" where electrons climb and descend, releasing light along the way.

3.8. ENERGY LEVEL DIAGRAMS AND TRANSITION NOTATIONS

Energy level diagrams and transition notations are highly effective atomic and quantum physics instruments. They play a crucial role in facilitating the visualization and comprehension of the energy configuration of atoms and the transitions that transpire between their discrete energy levels. The diagrams presented herein offer a lucid and concise depiction of electronic configurations, energy transitions, and the accompanying spectral lines discharged or assimilated by atomic entities. Energy level diagrams illustrate the organization of energy levels within an atom, visually representing the permissible electronic configurations for its constituent electrons. The diagrams presented herein visually represent the discrete energy levels accessible to electrons during their orbital motion around the atomic nucleus (Markeev et al., 2021). The y-axis of an energy level diagram commonly represents the principal quantum number (n), while the x-axis represents the electron subshells. The designation of each electron subshell is determined by the azimuthal quantum number (l) and the magnetic quantum number (ml), which define the orbital's shape and orientation.

An illustration of the energy level diagram for hydrogen, the most basic atom containing a solitary electron, comprises a sole energy level (n = 1) depicted along the y-axis. The horizontal axis represents the electron subshells s, p, d, and f. The electron within the hydrogen atom is located within the 1s orbital, representing the sole subshell accessible within the initial energy level.

As the atomic structure progresses towards atoms with multiple electrons, the energy level diagrams exhibit increased intricacy, characterized by the involvement of multiple energy levels and subshells. The number of electrons that can be accommodated in each energy level is determined by the $2n^2$ rule, where n represents the principal quantum number.

Transition notations are symbolic representations that depict the transitions between different energy levels within an atom. Utilizing these notations is crucial for comprehending the mechanisms involved in the emission and absorption of photons during electronic transitions. The prevalent method of denoting transitions typically employs the principal quantum numbers (n) to symbolize the initial and final energy levels implicated in the transition. An illustration of a transition from the energy level with principal quantum number $n = 3$ to the energy level with principal quantum number $n = 2$ is denoted as $3 \rightarrow 2$. Furthermore, to accommodate the various electron subshells and the permissible transitions between them, supplementary letters are employed in the notations for these transitions. The letters s, p, d, and f denote the electron subshells. An example of a transition from the 3p subshell to the 2s subshell can be represented as 3p $\rightarrow$ 2s (Eremin et al., 2011).

3.8.1. The Balmer Series

The Balmer series is widely recognized as an illustrative instance of energy level diagrams and transition notations, specifically in the context of hydrogen. The Balmer series pertains to the emission lines detected within the visible spectrum due to the transition of electrons in hydrogen atoms from higher energy levels to the second energy level ($n = 2$). The energy level diagram of hydrogen reveals that the first energy level ($n = 1$) is close to the nucleus and can accommodate a solitary electron within the 1s orbital. The second principal quantum level ($n = 2$) encompasses the 2s and 2p subshells, which can accommodate a maximum of eight electrons. When hydrogen electrons gain energy, they transition to higher energy levels (specifically, levels with principal quantum numbers greater than 2) in an excited state. Nevertheless, it should be noted that these higher energy states exhibit a lack of stability, leading to the eventual reversion of electrons back to the second energy level ($n = 2$). This process results in the emission of surplus energy in the form of photons (Onorato et al., 2015). The Balmer series exhibits several notable spectral lines, including the H-alpha line (transition from n = 3 to n = 2, 3 → 2) situated in the red region of the electromagnetic spectrum, the H-beta line (transition from n = 4 to n = 2, 4 → 2) found in the blue-green region, and the H-gamma line (transition from n=5 to n = 2, 5 → 2) observed in the violet region, among others. As the principal quantum number of the initial energy level increases, the Balmer series exhibits an extension into the ultraviolet region (Casasayas-Barris et al., 2019).

3.8.2. The Lyman Series and the Paschen Series

Hydrogen's emission spectrum exhibits additional spectral lines in conjunction with the Balmer series. The Lyman series illustrates electron transitions from higher energy levels to the first energy level, denoted by the principal quantum number n = 1. These transitions lead to the production of emission lines in the ultraviolet spectrum. The Lyman-alpha line, which corresponds to the transition from the second to the first energy level (2 → 1), is considered the most prominent line in the Lyman series. Conversely, the Paschen series encompasses electron transitions from energy levels of greater magnitude to the third energy level (n = 3). These transitions result in the emission of infrared spectral lines. The Paschen-alpha line, which corresponds to the transition from the fourth to the third energy level (4 → 3), is the most prominent in the Paschen series (Merrill & Wilson, 1934).

3.8.3. Fine Structure and Hyperfine Structure

The energy level diagrams and transition notations that have been previously presented provide a simplified representation of atomic spectra. However, it is important to note that the actual spectral lines exhibit greater complexity due to the presence of fine and hyperfine structures.

The phenomenon of fine structure is a consequence of the interplay between the intrinsic spin of an electron and its orbital angular momentum. These interactions result in minor variations in energy levels among sublevels within a given energy level. The phenomenon known as fine structure splitting is responsible for detecting multiple closely spaced lines in each spectral transition.

The hyperfine structure refers to a more refined level of energy splitting that emerges due to the interaction between the magnetic dipole moment of the electron and the nuclear magnetic dipole moment. The hyperfine structure introduces additional complexity to the spectral lines, resulting in supplementary fine characteristics (Axner et al., 2004).

3.8.4. Laser and Emission Spectroscopy Applications

Energy level diagrams and transition notations are paramount in comprehending and manipulating laser systems. Lasers, also known as light amplification by stimulated emission of radiation, operate based on the fundamental principles of atomic emission processes, enabling the generation of light beams that are both highly coherent and intense.

In the context of a laser, population inversion is realized, wherein the number of electrons occupying higher energy states surpasses the number of electrons residing in lower energy levels.

Stimulated emission is a phenomenon in which incident photons induce subsequent emissions in excited atoms, resulting in the coherent amplification of light. Consequently, the photons that are emitted exhibit coherent propagation, generating a concentrated and high-intensity laser beam.

Analyzing spectral lines emitted by atoms and molecules in emission spectroscopy relies heavily on utilizing energy level diagrams and transition notations. By examining spectral lines, researchers can discern the elemental composition, investigate the energy configuration of atoms and molecules, and acquire a deeper understanding of electron dynamics within these entities (Aragón & Aguilera, 2008).

3.9. BOHR MODEL AND ITS LIMITATIONS

The Bohr model of the atom, which was introduced by Niels Bohr in 1913, represented a notable advancement in comprehending the underlying composition and dynamics of atoms. This groundbreaking model played a pivotal role in explaining the emission spectrum of hydrogen and establishing a theoretical structure for quantizing energy levels within atomic systems.

The Bohr model of the atom constituted a pioneering theoretical framework that integrated principles of classical mechanics with components of the developing quantum theory.

The Bohr model postulated that electrons orbit the atomic nucleus in circular trajectories, analogous to the motion of *celestial* bodies around the sun. The electron orbits exhibited quantization, indicating that electrons were restricted to occupying distinct energy levels or shells. Electrons that are in closer proximity to the nucleus are situated within lower energy levels, whereas those that are located at greater distances occupy higher energy levels (Ginocchio & Kirson, 1980).

3.9.1. Explaining the Emission Spectrum of Hydrogen

The Bohr model demonstrated a notable accomplishment in explaining the emission spectrum of hydrogen. An element's emission spectrum comprises discrete lines of light that correspond to electron transitions occurring between different energy levels.

The Bohr model provided a concise and elegant explanation for electron transitions within hydrogen atoms, explaining the movement of electrons between distinct energy levels.

When an electron undergoes the absorption of energy and transitions to a higher energy level, it enters a state of excitation. Upon the electron's transition to a lower energy state, it emits the surplus energy as a photon, generating distinct spectral lines within the emission spectrum.

One notable illustration of this phenomenon is the Balmer series, which comprises a collection of distinct and luminous spectral lines within the visible spectrum. Hydrogen atoms emit these lines during the transition of their electrons from higher energy states to the second energy level ($n = 2$). The experimental validation of the Bohr model's predictions was verified, and observing these spectral lines further consolidated its significance (Cruzeiro et al., 2021).

3.9.2. The Quantization of Angular Momentum

The Bohr model identified the quantization of angular momentum as a significant revelation. Based on the principles of classical mechanics, it is postulated that an electron in orbit around the nucleus can occupy a continuum of paths, thereby giving rise to a continuous spectrum of potential energy values. Nonetheless, the Bohr model introduced the notion of quantized angular momentum, postulating that the angular momentum of an electron could solely assume discrete values determined by multiples of Planck's constant divided by 2π. The quantization process represented a significant departure from the principles of classical physics and served as a pivotal milestone in the advancement of quantum mechanics. The discovery was made that electrons within atoms can occupy distinct energy levels, laying the groundwork for comprehending the stability and electronic arrangements of atoms (Lloyd et al., 2012).

3.9.3. Limitations of the Bohr Model

Although the Bohr model provided significant contributions to the field, it was accompanied by several constraints that were subsequently resolved by more sophisticated theories, notably quantum mechanics. Several significant limitations are associated with the Bohr model which are discussed in subsections.

3.9.3.1. Failure to Explain Fine Structure

The Bohr model could not explain the intricate details observed in the fine structure of atomic spectral lines. In actuality, it was discovered that spectral lines exhibit a closely spaced arrangement consisting of multiple components. This observation suggests the existence of more refined energy disparities between electron sublevels within a given energy level. The limitations of the Bohr model in explaining fine structure phenomena necessitated the formulation of a more advanced theoretical framework capable of comprehensively addressing the intricate interplay between electrons and the atomic nucleus. This imperative ultimately led to the emergence and subsequent refinement of quantum mechanics.

3.9.3.2. Inability to Explain Hyperfine Structure

An additional constraint of the Bohr model pertained to its inadequacy in explaining hyperfine structure, which denotes a more intricate division observed in atomic spectral lines. The phenomenon of hyperfine structure is a consequence of the mutual interaction between an electron's magnetic dipole moment and the atomic nucleus's magnetic dipole moment.

The Bohr model's assumption of electrons as point-like particles devoid of intrinsic properties, such as spin, rendered it incapable of explaining the hyperfine structure phenomenon. The constraint above was successfully addressed through the advancement of quantum mechanics, which introduced the fundamental notion of electron spin.

3.9.3.3. Lack of Explanation for Electron's Wave Nature

The Bohr model postulated that electrons exhibit particle-like behavior by occupying discrete, well-defined orbits. Nevertheless, as the research advanced, it became apparent that electrons demonstrated wave-like characteristics, fundamental aspects of quantum mechanics. The electron's behavior within atoms can be more accurately represented by wavefunctions in quantum mechanics, which describe its wave-like nature. This wave-like nature allows a better understanding of the probabilities of finding electrons in specific regions around the nucleus.

3.9.3.4. The Quantum Revolution

Despite its inherent limitations, the Bohr model of the atom played a pivotal role in the advancement of quantum mechanics. The field of quantum mechanics, which was developed in the 1920s, presented a more comprehensive and precise theoretical framework for comprehending the behavior of atoms.

The discrete orbits of the Bohr model were superseded by wavefunctions in quantum mechanics, which describe the probability distribution of the electron's presence in various regions surrounding the nucleus. The concept of the Heisenberg uncertainty principle was introduced, which posits an inherent uncertainty in the simultaneous determination of a particle's position and momentum (McKagan et al., 2008).

3.10. QUANTUM MECHANICAL MODEL AND ATOMIC SPECTRA

The quantum mechanical model of the atom is a highly significant theoretical framework that has brought about a paradigm shift in the comprehension of atomic and subatomic phenomena. The model under discussion emerged during the initial decades of the 20th century, supplanting the classical conception of the atom. It introduced a probabilistic framework grounded in wavefunctions, enabling a more comprehensive and precise depiction of atomic spectra.

3.10.1. The Failure of Classical Physics

During the transition from the 19th to the 20th century, classical physics emerged as a highly effective framework for comprehending macroscopic phenomena, owing to its deterministic principles and the incorporation of continuous energy levels. Nevertheless, classical physics encountered substantial obstacles when applied to the atomic realm. The existing model did not adequately consider the empirical findings related to atomic spectra, specifically the fine structure and discrete spectral lines observed in the emission and absorption of light by atoms. The need for a novel theoretical framework became evident to reconcile the disparities between classical physics and atomic behavior. This event signified the initiation of the quantum revolution in the field of physics.

3.10.2. Wave-Particle Duality

The genesis of the quantum mechanical model emerged from the amalgamation of two conflicting notions: the wave-like characteristics exhibited by particles and the particle-like characteristics exhibited by waves, commonly referred to as wave-particle duality. The initial proposition of this concept can be attributed to Louis de Broglie in 1924, wherein he postulated the potential for particles, specifically electrons, to manifest wave-like characteristics.

The concept of wave-particle duality was a fundamental basis for advancing quantum mechanics, subsequently bringing about a transformative shift in the comprehension of atomic phenomena. The recognition that matter can exhibit wave-like and particle-like properties posed a significant challenge to classical physics and thus demanded the adoption of a novel framework (Zhu et al., 1995).

The quantum mechanics field, independently formulated by several prominent physicists such as Max Planck, Albert Einstein, Niels Bohr, Werner Heisenberg, and Erwin Schrödinger, introduced a groundbreaking theoretical framework for comprehending the behavior of atoms. The fundamental tenets of quantum mechanics encompass the subsections.

3.10.3. Wavefunctions and Probability

The wavefunction (ψ) is a mathematical function that describes the state of an electron within an atom in the context of quantum mechanics. The square of the magnitude of the wavefunction, denoted as ψ^2, provides the probability density distribution for locating the electron in proximity to the nucleus.

In contrast to classical physics, which assumes precise determination of particle position and momentum, quantum mechanics introduces inherent uncertainty. The Heisenberg uncertainty principle posits that the precise determination of a particle's position and momentum is inherently unattainable.

3.10.4. Quantized Energy Levels

The concept of quantized energy levels is widely regarded as the most groundbreaking aspect of quantum mechanics. In contrast to the principles of classical physics, which allow for the continuous distribution of energy levels, the framework of quantum mechanics asserts that electrons within atoms are restricted to occupying distinct and discrete energy levels or states.

The determination of quantized energy levels relies upon the solutions derived from the Schrödinger equation, a pivotal equation within the realm of quantum mechanics that characterizes wavefunctions' behavior when subjected to an atom's potential energy. Every energy level is associated with a distinct combination of quantum numbers that describe an electron's state.

3.10.5. Electron Orbitals and Atomic Spectra

The presence of quantized energy levels within the framework of the quantum mechanical model leads to the formation of electron orbitals, which delineate spatial domains wherein electrons exhibit the highest probability of being located near the atomic nucleus. The orbitals above possess distinctive geometric configuration, magnitude, and energy level attributes. The assignment of quantum numbers is employed to designate electron orbitals. The principal quantum number (n) indicates the primary energy level or shell occupied by an electron. The azimuthal quantum number (l) designates the subshell and ascertains the orbital's shape (s, p, d, f, etc.). The magnetic quantum number (ml) serves as an indicator of the spatial orientation of an orbital (Ballentine, 2014).

3.10.6. The Role of Quantum Mechanics in Atomic Spectra

The quantum mechanical model offers a comprehensive and precise comprehension of atomic spectra, encompassing both emission and absorption spectra.

3.10.6.1. Emission Spectra

Emission spectra are generated as a result of the emission of photons by atoms during the transition of electrons from energy levels of higher to lower energy. Every element possesses a distinct arrangement of energy levels, resulting in a unique pattern of emitted light known as its emission spectrum.

The emission of photons in the quantum mechanical model is intricately connected to the quantized energy differences between different energy levels. When an electron transitions from a state of higher to lower energy, it emits energy as a photon. This photon possesses a distinct wavelength and energy that corresponds to the disparity in energy between the two levels above.An element's emission spectrum comprises distinct, luminous lines, with each line corresponding to a particular electron transition. The lines above correlate directly with the atom's energy levels and electronic configuration. This concept can be illustrated in the hydrogen emission spectrum, specifically in the Balmer series, where electron transitions to the second energy level (n = 2) occur.

3.10.6.2. Absorption Spectra

Absorption spectra are detected when atoms undergo photon absorption, resulting in a transition from lower to higher energy levels. When a sample of atoms is subjected to a continuous light spectrum, certain wavelengths are preferentially absorbed, resulting in dark lines against a bright background.

In the context of the quantum mechanical model, absorption lines indicate the disparities in energy levels that electrons assimilate to undergo transitions to higher states. The absorption spectrum of an element is a distinct characteristic that provides insight into the permissible transitions of energy levels within the atom (Savall-Alemany et al., 2016).

3.10.6.3. Limitations and Advancements

The quantum mechanical model has significantly transformed the comprehension of atomic behavior and spectra; however, it is not without its constraints. The complexity of the quantum mechanical model escalates when considering atoms with multiple electrons. The computational challenges arising from the interactions among electrons and their mutual repulsion necessitate the utilization of approximations and advanced mathematical techniques.

1. **Relativistic Effects:** The quantum mechanical model fails to incorporate relativistic effects, which become prominent in elements with large atomic nuclei, particularly those located in the inner sections of the periodic table. The accurate depiction of the behavior of these elements necessitates the utilization of relativistic quantum mechanics or relativistic quantum field theory (QFT).

2. **Quantum Field Theory (QFT):** In contexts characterized by exceedingly high energy levels, such as particle accelerators or the nascent stages of the universe, the theoretical framework of quantum mechanics necessitates expansion to encompass the tenets of QFT. QFT serves as a unifying framework that combines the principles of quantum mechanics and special relativity, providing a comprehensive description of particles as manifestations of quantum field excitations.

The quantum mechanical model of the atom represents a significant milestone in the field of physics, as it challenges the conventional perspective of the atomic world and facilitates a more comprehensive understanding of atomic spectra and behavior. The theoretical framework presented in this model is grounded in the fundamental concepts of wave-particle duality, wavefunctions, and quantized energy levels. It offers a robust and comprehensive approach to understanding the intricate dynamics of the subatomic domain (Rezac & Hobza, 2012).

3.11. SPECTROSCOPIC TECHNIQUE

Spectroscopic techniques have become prominent tools that facilitate the investigation and examination of the characteristics of substances by examining the interaction between light and various materials. By examining their interactions with light, Spectroscopy offers valuable insights into the structure, composition, and behavior of atoms, molecules, and materials.

The field of spectroscopy is based on the fundamental principles governing the interaction between light and matter. When light traverses or interacts with a substance, it experiences distinct alterations that convey significant insights into its properties. The fundamental principle underlying spectroscopy is predicated upon the notion that atoms and molecules exhibit distinct energy levels. Consequently, when these entities absorb or emit light, they exclusively do so at precise wavelengths corresponding to these particular energy transitions (Newman et al., 1986).

3.11.1. Absorption Spectroscopy

Absorption spectroscopy is considered a fundamental and extensively employed technique within the field. The process entails the measurement of the light wavelengths that are absorbed by a given sample. When electromagnetic radiation traverses a substance, some of the photons are assimilated by the constituent atoms or molecules, thereby causing the promotion of their associated electrons to states of increased energy. The energy absorbed is directly associated with particular transitions occurring between the energy levels of the material.

Absorption spectroscopy is utilized across various scientific fields, encompassing analytical chemistry, environmental monitoring, and pharmaceutical analysis. Ultraviolet-visible (UV-Vis) spectroscopy is a widely employed technique that utilizes light within the ultraviolet and visible regions of the electromagnetic spectrum (Wende, 2004).

3.11.2. Emission Spectroscopy

Emission spectroscopy is a scientific technique that quantifies the electromagnetic radiation emitted by a substance after its stimulation by external energy input. When atoms or molecules within a substance assimilate energy, their electrons transition to elevated energy states. Upon the reversion of electrons to their lowest energy level, they discharge the surplus energy by emitting photons, thereby generating light with distinct wavelengths.

Atomic emission spectroscopy (AES) is an exemplary illustration of this analytical technique. The characteristic wavelengths of emitted photons allow for identifying and quantifying elements within a given material, thus facilitating the analysis conducted by scientists. The emission spectrum is a distinctive characteristic that provides insight into the permissible transitions of energy levels within atoms.

The equation that describes the energy of a photon thus emitted is:

$$E = hc/\lambda$$

The equation provided relates the energy of a photon (E) to Planck's constant (h), the speed of light (c), and the wavelength of the emitted light (λ). Emission spectroscopy finds extensive utility across diverse disciplines, notably in astrophysics, where it serves as a valuable tool for investigating the elemental composition of stars and interstellar gas (Barnett et al., 1968).

3.11.3. Scattering Spectroscopy

Scattering spectroscopy examines light that undergoes deflection or scattering when interacting with a given specimen. When incident light interacts with particles or molecules within a material, it changes direction due to different scattering mechanisms, including Rayleigh scattering and Raman scattering. Rayleigh scattering is a phenomenon that takes place when light is dispersed by particles or molecules that possess dimensions significantly smaller than the wavelength of the incident light. This phenomenon is accountable for the azure sky's hue during daylight hours.

In contrast, Raman scattering is characterized by inelastic scattering, wherein the energy of the scattered light deviates from that of the incident light. Raman spectroscopy is a technique that enables the determination of molecular vibrational and rotational energy levels, facilitating the identification of chemical compounds and the analysis of molecular structures. Using scattering spectroscopy is of utmost importance in material characterization, as it offers valuable insights into the molecular structure and dynamics of the sample under investigation (Bigio & Mourant, 1997).

3.11.4. Nuclear Magnetic Resonance (NMR) Spectroscopy

Nuclear magnetic resonance (NMR) was first detected experimentally at the end of 1945, nearly concurrently with the work groups Felix Bloch, Stanford University, and Edward Purcell, Harvard University. The first NMR spectrum was first published in the same issue of the Physical Review in January 1946. Bloch and Purcell were jointly awarded the 1952 Nobel Prize in Physics for their NMR spectroscopy research.

NMR spectroscopy is a crucial analytical tool for organic chemists. The research in the organic lab has been significantly improved with the aid of the NMR. Not only can it provide information on the molecule's structure, but it can also determine the content and purity of the sample. Proton (1H) NMR is one of organic chemists most widely used NMR methods. The protons present in the molecule will behave differently depending on the surrounding chemical environment, making it possible to discuss their structure (Gerothanassis et al., 2002).

3.11.4.1. Principle

According to the principles of NMR, it is observed that a significant number of nuclei possess spin while simultaneously exhibiting electrical charge. The transfer of energy from a lower to a higher can be accomplished by providing an external magnetic field.

- All atomic nuclei possess electrical charge and a significant number of them exhibit spin.

- Applying an external magnetic field enables energy transfer from a lower energy state to higher energy levels.
- The energy transmission occurs at a wavelength that aligns with the radio frequency (RF).
- Additionally, energy emission occurs at an identical frequency when the spin returns to its fundamental state.
- Hence, the processing of the NMR spectrum for the specific nucleus is achieved by measuring the corresponding transfer signal.

3.11.4.2. Working

The sample should be subjected to a magnetic field. The nuclei sample is stimulated into a state of NMR by applying radio waves, generating NMR signals. The NMR signals are detected using highly sensitive radio receivers. The intramolecular magnetic field surrounding an atom in a molecule induces a change in its resonance frequency. This provides a comprehensive analysis of the distinct functional groups present in a molecule and an in-depth examination of its electronic configuration. NMR spectroscopy is a highly reliable technique for definitively identifying monomolecular organic compounds. This methodology offers comprehensive information regarding a given molecule's reaction state, molecular structure, chemical environment, and molecular dynamics (Antequera et al., 2021).

3.11.4.3. Chemical Shift in NMR

The phenomenon of a spinning charge giving rise to a magnetic field is accompanied by generating a magnetic moment directly proportional to the spin. When subjected to an external magnetic field, two distinct spin states can be observed on a charge: spin up and spin down. In the spin-up state, the spin aligns with the direction of the magnetic field, while in the spin-down state, the spin opposes the magnetic field. The chemical shift is the difference between the resonant frequency of spinning protons and the frequency emitted by the reference molecule. In the phenomenon of NMR, chemical change holds significant importance in molecular structure determination. NMR spectroscopy allows for the detection of various nuclei, including 1H (proton), 13C (carbon-13), 15N (nitrogen-15), and 19F (fluorine-19), among others. The isotopes 1H and 13C are extensively employed in various applications. The definition of 1H is highly descriptive in the context of NMR spectroscopy. The study of mechanics reveals that the motion of an electric charge gives rise to the generation of a magnetic field. In the context of NMR, applying radio frequency (RF) radiation to the nucleus induces rotational motion of the nucleus and its associated magnetic field. (Govindaraju et al., 2000).

3.11.4.4. NMR Spectroscopy Techniques

1. **Resonant Frequency:** The term pertains to the energy associated with the absorption process, while the signal's intensity is directly proportional to the magnitude of the magnetic field. Nuclei that exhibit NMR activity can absorb electromagnetic radiation at a frequency specific to the isotope when subjected to a magnetic field.
2. **Acquisition of Spectra:** An NMR response is obtained when the sample is excited with a radiofrequency pulse. This signal exhibits a notably low strength, necessitating the utilization of highly sensitive radio receivers for its detection (Meyer & Peters, 2003) (Figure 3.6).

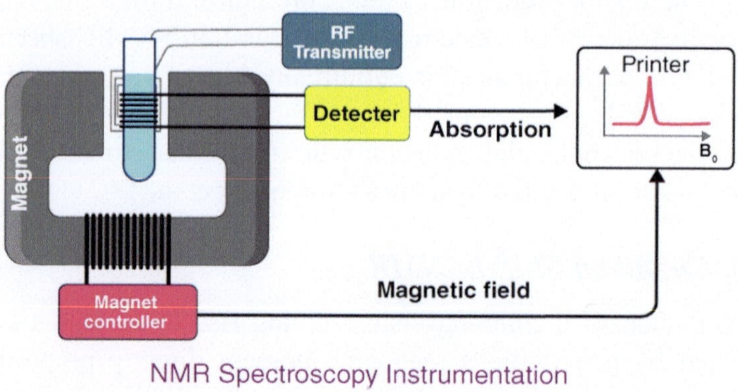

Figure 3.6. *NMR spectroscopy instrumentation.*

Source: Calvinn, Creative Commons License.

3.11.5. Mass Spectrometry

Mass spectrometry is a highly adaptable methodology employed to ascertain the mass and composition of molecules. Mass spectrometry involves the ionization of a given sample, followed by the subsequent separation of the resulting ions according to their mass-to-charge ratio (m/z) (Figure 3.7).

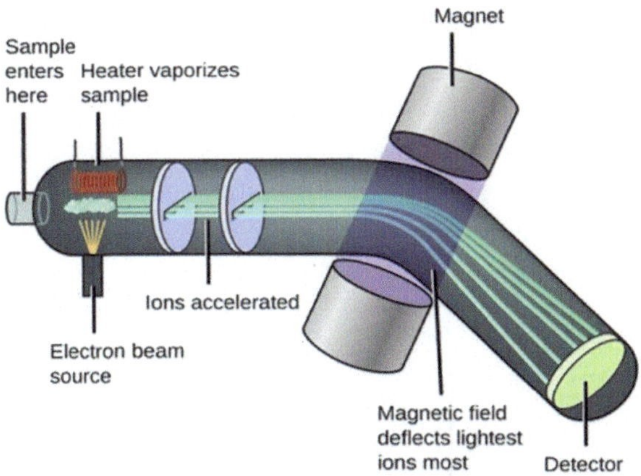

Figure 3.7. *Scheme of the mass spectrometry process.*

Source: Preedy, Creative Commons License.

1. **Ionization Source:** Molecules are converted into gas-phase ions, enabling their mobility and manipulation through the application of external electric and magnetic fields. The laboratory employs a technique known as nanoelectrospray ionization, which resembles the industrial process of automotive painting. This methodology facilitates the generation of ions with either positive or negative charges, contingent upon the specific demands of the experiment. Nanoelectrospray ionization enables the direct connection between the outlet of a reduced-scale chromatography column and a mass spectrometer's inlet. The fluid emanating from the column is directed through a needle with a diameter ranging from 10 to 15 micrometers at its apex (Mann et al., 2001).
2. **Mass Analyzer:** After ionization, the ions are subsequently sorted and separated based on their mass-to-charge (m/z) ratios. Various mass analyzers are presently accessible, each exhibiting distinct trade-offs pertaining to operational speed, separation resolution, and other operational prerequisites. The mass analyzer frequently operates in conjunction with the ion detection system.
3. **Ion Detection System:** Subsequently, the ions that have been isolated are quantified and transmitted to a data management system, wherein the mass-to-charge (m/z) ratios are recorded alongside their corresponding relative abundances. The mass spectrum is a graphical representation of the relative intensities of

ions in a sample, plotted against their mass-to-charge ratios (m/z). Every peak observed in a mass spectrum represents a distinct component with a unique mass-to-charge ratio (m/z) within the analyzed sample. The relative abundance of these components can be inferred from the heights of the peaks (Ruhaak et al., 2018) (Figure 3.8).

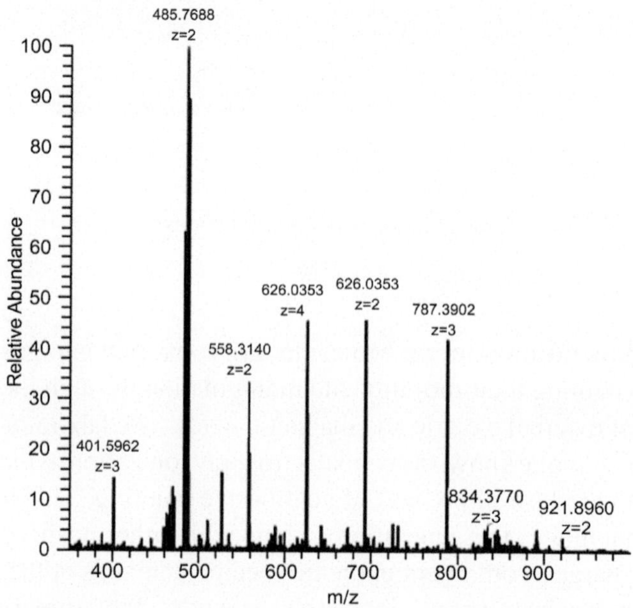

Figure 3.8. *Example of the mass spectrum.*

Source: P. J. Potts, Creative Commons License.

3.11.6. X-Ray Spectroscopy

X-ray spectroscopy is a highly effective methodology for investigating various substances' electronic configurations. The phenomenon under consideration encompasses the interaction between X-rays and matter, leading to various spectroscopic effects, including X-ray emission, absorption, and scattering. X-ray absorption spectroscopy (XAS) is a technique that offers insights into the electronic structure of atoms within a given material, thereby enabling the determination of the oxidation state and local coordination environment (Peterson & Fabian, 2006).

3.11.7. Infrared Spectroscopy

The infrared spectroscopy technique is extensively employed to examine molecular vibrational modes. Molecules absorb infrared radiation, leading to vibrational transitions between distinct energy states.The infrared spectrum exhibits a succession of peaks, each corresponding to a distinct vibrational mode of the molecule. In the field of organic chemistry, the utilization of infrared spectroscopy is widespread. It determines molecular structure, identifies functional groups, and characterizes polymers (Stuart, 2000; Wang et al., 2017).

3.12. ATOMIC SPECTROSCOPY

Electrons are present in discrete energy levels within the atomic structure, commonly referred to as atomic orbitals. Atomic orbitals exhibit quantization, implying that they possess discrete, well-defined values rather than a continuous spectrum.

The movement of electrons between orbitals necessitates the absorption or emission of energy equivalent to the disparity in energy levels among the quantized orbital states of their respective atoms. In optical spectroscopy, the absorption or emission of energy associated with the transition of an electron from a lower energy level (lower orbital) to a higher energy level (higher orbital) is manifested through the absorption or emission of photons, which are discrete particles of light.

Due to the distinct number of electrons associated with each element, atoms exhibit a characteristic energy absorption or emission pattern specific to their elemental identity (e.g., Ca, Na, etc.). Consequently, these atoms absorb or emit photons in a correspondingly distinctive pattern. Determining the composition or quantity of atoms in a given sample can be achieved by analyzing alterations in both the wavelength and intensity of light (Bings et al., 2010) (Figure 3.9).

Atomic spectroscopy can be classified into three main categories: atomic absorption, mass spectroscopy, and emission spectroscopy. Atomic absorption spectroscopy (AAS) involves the transmission of light with a specific wavelength through a group of atoms. When the wavelength of the incident light matches the energy difference between two energy levels in the atoms, a fraction of the light will undergo absorption.

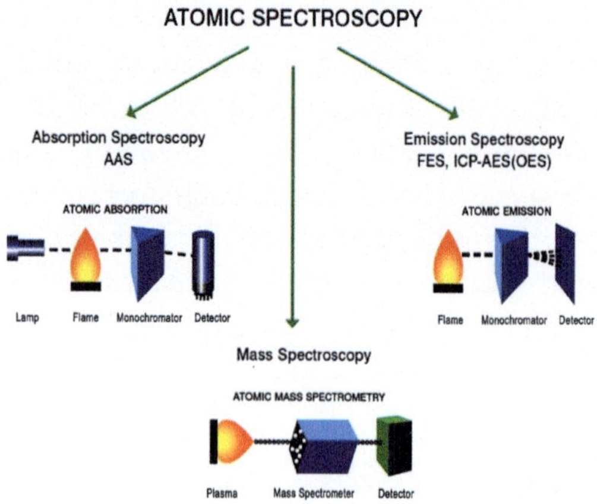

Figure 3.9. *Types of atomic spectroscopy.*

Source: AWE Magazine, Creative Commons License.

The difference in the magnitude of the luminous output originating from the light source, such as a lamp, and the luminous input captured by the detector results in the determination of an absorbance measurement. The obtained absorbance value can subsequently be utilized to quantify the concentration of a specific element (or atoms) in the sample. The Beer-Lambert law provides a mathematical representation of the correlation between the concentration of atoms, the distance traversed by light through a collection of atoms, and the extent to which the light is absorbed. In the context of AES, it can be observed that the intensity of the emitted light exhibits a direct proportionality to the concentration of atoms (Rudzikas, 2007).

3.12.1. Atomic Emission Spectroscopy (AES)

Atomic emission spectroscopy (AES) is a scientific technique employed for chemical analysis, wherein the amount of an element present in a sample is determined by measuring the intensity of light emitted at a specific wavelength from a flame, plasma, arc, or spark. The determination of an element's identity in the emission spectrum is facilitated by the wavelength of its atomic spectral line. In contrast, the quantity of emitted light is directly proportional to the abundance of atoms belonging to the element. The sample can be stimulated through a variety of methods (Figure 3.10).

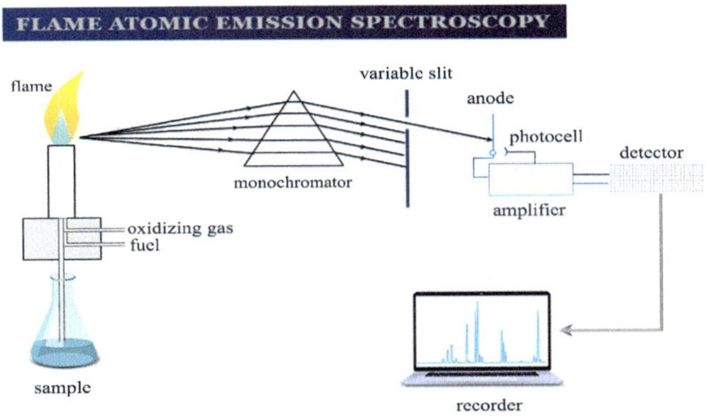

Figure 3.10. *Schematic illustration of atomic emission spectroscopy.*

Source: Learning Chemistry, Creative Commons License.

AES is a highly effective analytical methodology employed for identifying and quantifying elements within a given sample, relying on the distinct wavelengths of light emitted by atoms in an excited state. This method exhibits significant value in many scientific and industrial contexts, encompassing environmental monitoring, metallurgy, forensic analysis, and other related domains (Uden, 1992).

3.12.1.1. The Basis of Atomic Emission Spectroscopy (AES)

Atomic emission spectroscopy is based on the fundamental principle that when atoms are stimulated, usually by introducing energy in the form of light, heat, or electricity, the electrons within them transition from lower energy states to higher energy states. Upon their transition back to the ground state, these electrons emit surplus energy in the form of photons, which are discrete units of electromagnetic radiation. AES is a dependable technique for identifying elements due to the distinctive wavelengths exhibited by the emitted photons.

3.12.1.2. Instrumentation and Experimental Setup

The typical experimental configuration for AES comprises a spectrometer, a light source utilized for atoms' excitation, and a sample introduction system. The utilization of a spectrometer is imperative in the process of separating the light that is released into its constituent wavelengths and subsequently measuring its magnitude.

Utilizing a hollow cathode lamp as the primary light source is a prevalent practice in AES. The hollow cathode lamp is equipped with a cathode composed of the specific element under investigation, which emits light upon passing a discharge current. The light emitted comprises distinct spectral lines corresponding to the specific electronic transitions occurring within the excited atoms (Shkirskiy et al., 2015).

3.12.1.3. Spectral Analysis and Wavelength Selection

The light emitted by the sample consists of distinct lines, with each line corresponding to a particular electron transition occurring within the excited atoms. The spectrometer is responsible for the dispersion of light into its constituent wavelengths, while a detector is employed to quantify the intensity of each wavelength. The spectrum that is produced as a result is commonly referred to as the atomic emission spectrum.

It is imperative to carefully choose suitable wavelengths for the analysis to achieve precise identification and quantification of the elements present in the sample. The process is commonly conducted by comparing the detected emission lines with a comprehensive collection of established atomic emission spectra about different elements. Every individual element possesses a distinct and exclusive emission line collection that serves as its distinctive spectral signature (Middleton & Price, 1996).

3.12.1.4. Calibration and Quantitative Analysis

Calibration curves are constructed to conduct quantitative analysis utilizing AES. A set of standardized solutions containing predetermined concentrations of the element under investigation is prepared and analyzed using identical experimental parameters employed for the sample. The intensities of the emission lines are measured for each standard solution, and calibration curves are constructed by graphing the intensity as a function of the concentration. After establishing the calibration curves, the measurement of emission line intensities in the sample is conducted, enabling the determination of the element's concentration in the sample through interpolating from the calibration curve (Bersier et al., 1994).

3.12.1.5. Applications of Atomic Emission Spectroscopy (AES)

Atomic emission spectroscopy (AES) is extensively utilized across diverse disciplines due to its broad scope of applications. Applying AES in environmental analysis enables identifying and quantifying minute quantities of elements present in water, soil, and air samples. This analytical technique is crucial in monitoring the environment and pollution control efforts.

In the domain of metallurgy, AES plays a pivotal role in ensuring quality control by enabling the examination of the elemental makeup of metal alloys and the detection of impurities.

AES is a valuable technique that contributes to forensic analysis. It plays a crucial role in identifying various types of trace evidence, such as gunshot residues and paint fragments.

AES, an acronym for Analytical Electrochemistry Spectroscopy, finds extensive application in the pharmaceutical sector for drug development and quality control. Its primary objective is to ascertain that medications adhere to predetermined specifications by ensuring the presence of specified elements within acceptable safety thresholds (Wienke et al., 1992).

3.12.1.6. Advantages and Limitations

AES provides numerous advantages. The technique is characterized by its high speed and sensitivity, enabling the detection of minute quantities of trace elements in intricate sample matrices. Moreover, this technique necessitates minimal sample preparation, rendering it relatively uncomplicated to execute.

Nevertheless, like other analytical techniques, AES has certain limitations. The applicability of this method is limited to elements that can be excited and emit light. Consequently, elements with high ionization energies or low vaporization temperatures might not be suitable for analysis using this technique. Furthermore, this technique is limited to identifying specific elements and does not yield information regarding molecular species (Fujihara & Nishimoto, 2020).

3.12.2. Atomic Absorption Spectroscopy (AAS)

Atomic absorption spectroscopy (AAS) and AES are electroanalytical techniques utilized for quantitatively analyzing chemical elements in their gaseous state, specifically by measuring the absorption and emission of electromagnetic radiation by free atoms. AAS is a scientific technique that relies on light absorption by unbound metallic ions.

Analytical chemistry employs a methodology for ascertaining the concentration of a specific element, referred to as the analyte, within a sample designated for analysis. AAS is a widely employed analytical technique for determining more than 70 distinct elements in solution. Additionally, AAS can directly analyze solid samples through electrothermal vaporization. This technique finds applications in various fields of study, including pharmacology, biophysics, archaeology, and toxicology research.

The application of AES as an analytical method and the establishment of its fundamental principles can be attributed to the pioneering work of Robert Wilhelm Bunsen and Gustav Robert Kirchhoff, esteemed professors at the University of Heidelberg in Germany during the latter half of the 19th century (Penache et al., 2002).

The contemporary iteration of AAS was predominantly formulated in the 1950s by a group of chemists hailing from Australia. The research team was under the guidance of Sir Alan Walsh, who held a position at the Commonwealth Scientific and Industrial Research Organization (CSIRO), specifically within the Division of Chemical Physics, located in Melbourne, Australia.

Atomic absorption spectrometry (AAS) finds extensive applications across various domains of chemistry, including the clinical analysis of metals present in biological fluids and tissues. These biological matrices encompass many samples, such as whole blood, plasma, urine, saliva, brain tissue, liver, hair, and muscle tissue. Atomic absorption spectrometry is a versatile analytical technique useful in qualitative and quantitative analyses (Christian, 1969) (Figure 3.11).

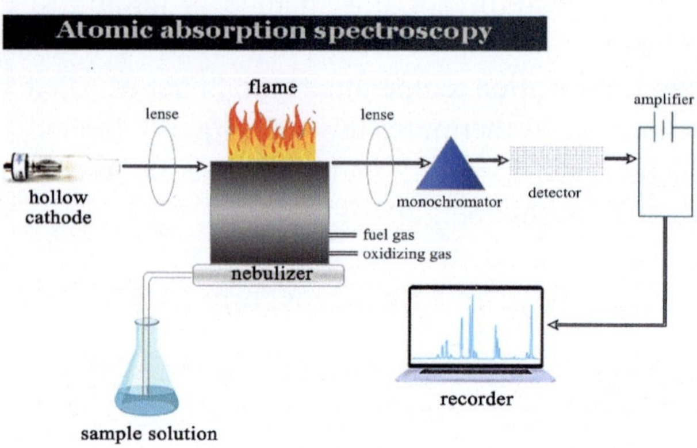

Figure 3.11. *Scheme of the process of atomic absorption analysis.*

Source: Uden, Creative Commons License.

3.12.2.1. Principles

The methodology utilizes the atomic absorption spectrum of a given sample to evaluate the concentration of particular analytes present within it. Establishing the relationship between the measured absorbance and the analyte concentration in a given sample necessitates the utilization of standards containing known analyte content. This process is reliant upon the principles outlined in the Beer-Lambert law.

3.12.2.2. Instrumentation

To analyze the atomic constituents, present in a sample, it is necessary to subject the sample to the atomization process. Today's prevailing atomizers encompass flame and electrothermal (graphite tube) atomizers. Subsequently, subjecting the atoms to optical radiation is imperative, which can be achieved using either an element-specific line radiation source or a continuum radiation source. Subsequently, the radiation traverses a monochromator to effectively segregate the element-specific radiation from any extraneous radiation emitted by the radiation source. A detector ultimately quantifies this isolated radiation.

3.12.2.3. Atomizers

The techniques commonly employed in contemporary research and analysis involve the utilization of spectroscopic flames and electro-thermal atomizers. Alternative atomization techniques may be employed for specific applications, such as hydride atomization, glow-discharge atomization, or cold-vapor atomization (David, 1958).

3.12.2.4. Flame Atomizers

The flame atomizers, specifically the air-acetylene flame with a temperature of approximately 2,300°C and the nitrous oxide-acetylene flame with a temperature of around 2,700°C, are the most ancient and widely employed in AAS. Moreover, the latter flame provides a more conducive atmosphere, making it particularly suitable for analytes with a strong affinity for oxygen.

Flame atomizers are commonly employed for the analysis of liquid or dissolved samples. The provided solution is subjected to aspiration using a pneumatic analytical nebulizer. This process converts the solution into an aerosol and then introduces it into a spray chamber. The aerosol is combined with the flame gases within the spray chamber and undergoes conditioning.

This conditioning ensures that only the smallest aerosol droplets, measuring less than 10 μm in diameter, can enter the flame. The conditioning above procedure mitigates interference; however, only approximately 5% of the aerosolized solution can reach the flame due to this process.

A burner head is located above the spray chamber, which generates a flame that possesses a significant lateral extent, typically ranging from 5 to 10 centimeters, while maintaining a shallow depth of only a few millimeters. The radiation beam traverses the flame along its longest axis, while the flow rates of the flame gas can be manipulated to achieve optimal levels of free atom concentration. The height of the burner can also be modified to ensure that the radiation beam traverses the region within the flame where the atom cloud density is at its maximum, thereby leading to the attainment of optimal sensitivity (Porra et al., 1989).

The flame undergoes several distinct processes: desolvation, vaporization, atomization, and ionization. Desolvation refers to solvent removal through evaporation, leaving behind dry nanoparticles of the sample. Vaporization involves the conversion of solid particles into gaseous molecules, while atomization dissociates molecules into individual free atoms. Lastly, ionization occurs when analyte atoms, depending on their ionization potential and the energy present in the flame, may be partially converted into gaseous ions.

Each stage presents the potential for interference if there is a discrepancy in the degree of phase transfer between the analyte in the calibration standard and the analyte in the sample. The ionization process is typically unfavorable due to its tendency to decrease the number of measurable atoms, thereby diminishing sensitivity.

A consistent and stable signal is produced during the aspiration of the sample in flame AAS. This methodology is commonly employed for quantifications within the milligrams per liter (mg/L) range. It can potentially be extrapolated to concentrations as low as a few micrograms per liter (μg/L) for certain elements (Maurya et al., 2018).

3.12.2.5. Electrothermal Atomizers

The electrothermal atomic absorption spectroscopy (ET AAS) technique utilizing graphite tube atomizers was first developed by Boris V. L'vov at the Saint Petersburg Polytechnical Institute in Russia during the latter part of the 1950s. Concurrently, Hans Massmann at the Institute of Spectrochemistry and Applied Spectroscopy (ISAS) in Dortmund, Germany, also investigated this method. Numerous graphite tube configurations have been employed throughout the years; however, contemporary dimensions generally adhere to a length range of 20–25 mm and an inner diameter range of 5–6 mm. This technique directly analyzes liquid, solid, and gaseous samples. A predetermined volume of a solid sample is introduced into the graphite tube, usually ranging from 10 to 50 μL, or a mass that is weighed, typically around 1 mg. The sample is then subjected to a temperature program. The process typically involves several stages, including drying, where the solvent is evaporated; pyrolysis, where most of the matrix constituents are eliminated; atomization, where the analyte element is released into the gaseous phase; and cleaning, where any remaining residues in the graphite tube are eliminated through high-temperature treatment (McNally & Holcombe, 1987).

The graphite tubes are subjected to heating through their ohmic resistance, employing a power supply characterized by low voltage and high current. This heating process enables precise control over the temperature in each stage, while the implementation of temperature ramps between these stages' aids in the separation of components within the sample. Tubes can be heated either transversely or longitudinally, with the former offering a more uniform temperature distribution along their entire length. The concept of the stabilized temperature platform furnace (STPF), as introduced by Walter Slavin and built upon the research of Boris L'vov, effectively eliminates interference in the ET AAS technique. The primary constituents of this concept encompass the fragmentation of the sample from a graphite platform, specifically the L'vov platform, which is inserted into the graphite tube. This approach is employed to defer the atomization process until the gas phase within the atomizer has attained a consistent temperature. Additionally, a chemical modifier is utilized to stabilize the analyte at a pyrolysis temperature that is adequate for eliminating most of the matrix components. Lastly, instead of relying on the absorbance peak height for quantification, here, the absorbance is integrated throughout the transient absorption signal.

In ET AAS, a transient signal is produced, and the magnitude of this signal is directly correlated with the mass of the analyte introduced into the graphite tube rather than its concentration. This technique's notable advantage is its ability to directly analyze samples of various states, including solids, liquids, and gases. The sensitivity of this method is significantly greater than that of flame AAS by a factor of 2–3 orders of magnitude. As a result, it is capable of accurately determining concentrations in the low microgram per liter ($\mu g\ L^{-1}$) range for a typical sample volume of 20 microliters (μL), as well as in the nanogram per gram (ng g^{-1}) range for a typical sample mass of 1 milligram (mg). The exceptional freedom from interferences exhibited by ET AAS positions it as the most resilient technique presently accessible for quantifying trace elements in intricate matrices.

3.13. SPECIALIZED ATOMIZATION TECHNIQUES

Although flame and electrothermal vaporizers are widely employed as atomization techniques, several other atomization methods are utilized for specific applications.

3.13.1. Glow Discharge Atomization

The glow-discharge device (GD) functions as a multifunctional source, which can simultaneously introduce and atomize the sample. Glow discharge occurs within a gaseous environment containing argon at a low-pressure range of 1 to 10 torr. Within this particular environment exists a set of electrodes that administer a direct current (DC) voltage ranging from 250 to 1,000 V. The purpose of this voltage is to induce the breakdown of argon gas into ions carrying positive charges and free electrons. When subjected to the electric field, the ions experience acceleration towards the cathode surface that houses the sample. This bombardment of the sample by the ions leads to the ejection of neutral sample atoms, a phenomenon referred to as sputtering. The atomic vapor generated through this discharge consists of ions, atoms in their ground state, and some in an excited state. The emission of a low-intensity glow, which gives the technique its name, occurs when the excited atoms return to their ground state.

One essential criterion for glow discharge atomizers is that they possess electrical conductivity. As a result, atomizers are predominantly utilized in examining metals and other samples with conducting properties. Nevertheless, through appropriate adjustments, it can be employed for the analysis of liquid samples and nonconducting materials by incorporating a conductor, such as graphite, into the mixture (Bendicho & de Loos-Vollebregt, 1991).

3.13.2. Hydride Atomization

Hydride generation techniques are specifically designed for the analysis of elemental solutions. The methodology offers a mechanism for introducing samples that contain arsenic, selenium, antimony, bismuth, and lead into an atomizer in the gaseous state. The utilization of hydride atomization in conjunction with these components results in a significant improvement in detection limits, ranging from 10 to 100 times higher compared to alternative methodologies. Hydride generation is achieved by introducing an acidified water solution of the sample into a glass vessel containing a 1% aqueous solution of sodium borohydride.

The volatile hydride produced due to the chemical reaction is transported into the atomization chamber through an inert gas, where it subsequently undergoes decomposition. The procedure above facilitates the generation of a fragmented state of the substance under analysis, enabling subsequent quantification through absorption or emission spectrometry techniques (Vil'pan et al., 2005).

3.13.3. Cold Vapor Atomization

The cold-vapor technique is a method of atomization that is exclusively applicable to the determination of mercury. Mercury is the only metallic element exhibiting significant vapor pressure at standard ambient temperature. Due to this characteristic, it is useful in analyzing organic mercury compounds within samples and explaining their environmental distribution.

The procedure involves converting mercury to Hg^{2+} through nitric and sulfuric acid oxidation. This is followed by reducing Hg^{2+} using tin(II) chloride. The mercury is subsequently transferred into a long-pass absorption tube by introducing an inert gas stream into the reaction mixture.

The gas concentration is determined by measuring its absorbance at a wavelength of 253.7 nm. The technique exhibits detection limits within the parts-per-billion range, thereby establishing it as a highly proficient method for atomization-based mercury detection (Grinshtein et al., 2001).

3.13.4. Radiation Source

It is imperative to differentiate between line source atomic absorption spectroscopy (LS AAS) and continuum source atomic absorption spectroscopy (CS AAS). In the classical technique of line-source AAS, as originally suggested by Alan Walsh, the essential high spectral resolution needed for accurate measurements is achieved through the inherent emission spectrum of the radiation source (Falk, 1991). This emission spectrum consists of narrower lines than the analyte's absorbed lines. Continuum sources, such as deuterium lamps, are exclusively employed for background correction.

One notable benefit of employing this methodology is the reduced requirement for a medium-resolution monochromator to measure AAS. However, a notable drawback is the typical necessity of utilizing individual lamps for each element that needs to be quantified. In computer science, specifically in AAS, a solitary lamp is employed to emit a continuous spectrum encompassing the entirety of the spectral range relevant to the analysis of all elements (Ozbek & Akman, 2012).

3.13.5. Hollow Cathode Lamps

Hollow cathode lamps (HCLs) are widely utilized as the predominant radiation source in Laser-Induced Breakdown Spectroscopy (LIBS) and AAS. The interior of the sealed lamp is occupied by a specific gas, such as argon or neon, maintained at a reduced pressure. Within this environment, a cylindrical metallic cathode, housing the element of interest, coexists with an anode. Applying a high voltage between the anode and cathode leads to the ionization of the fill gas (Chakrabarti et al., 1993).

The gas ions undergo acceleration towards the cathode. Upon collision with the cathode, they cause the ejection of the cathode material. This cathode material becomes excited in the glow discharge and subsequently emits radiation specific to the sputtered material, corresponding to the element of interest. In most instances, single-element lamps are employed, wherein the cathode is formed by pressing predominantly compound materials of the target element. Lamps containing multiple elements can be obtained by incorporating combinations of compounds of the desired elements into the cathode through a pressing process.

Multi-element lamps exhibit a slightly lower sensitivity level than their single-element counterparts, necessitating careful selection of element combinations to mitigate potential spectral interferences. Many multi-element lamps typically incorporate a few elements, ranging from 2 to 8. AAS is a widely used analytical technique in chemistry. It is employed to determine the spectrometers can be equipped with a minimal number of hollow cathode lamp positions, ranging from 1 to 2. Alternatively, in the case of automated multi-element spectrometers, a greater number of lamp positions, typically ranging from 8 to 12, may be available (Gilmutdinov et al., 1995) (Figure 3.12).

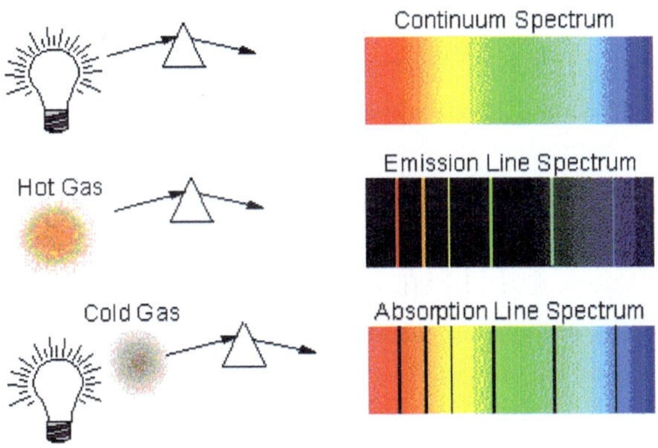

Figure 3.12. *Production of continuous spectrum.*

Source: Reynolds, Creative Commons License.

3.13.6. Electrodeless Discharge Lamps (EDLs)

Electrodeless discharge lamps (EDLs) comprise a quartz bulb containing a small amount of the analyte in the form of a metal or salt, along with an inert gas, typically argon, at low pressure.

These lamps are placed within a coil that generates an electromagnetic RF field, inducing a low-pressure inductively connected discharge within the lamp. Compared to hollow cathode lamps (HCLs), EDLs exhibit higher emission levels and generally possess narrower line widths. However, EDLs require a distinct power supply and may necessitate a longer stabilization period (Tittarelli et al., 1994).

3.13.7. Deuterium Lamps

Deuterium HCL, hydrogen HCL, and deuterium discharge lamps are employed in Laser-Induced Breakdown Spectroscopy (LIBS) for background correction. The emitted radiation intensity of these lamps exhibits a notable decline as the wavelength increases, limiting their applicability to the range of approximately 190 to 320 nm (Yuzefovsky et al., 1994).

Did You Know?

Infrared spectroscopy is used to analyze the breath of astronauts in space to monitor their health and detect any potential illnesses.

3.14. SUMMARY

The chapter discussed the fundamental concepts of atomic spectra and emissions, interpreting the complexities of interactions between light and matter and their profound implications within the domains of physics and chemistry.

The exploration commenced with examining the notion of atomic spectra, which serves as the fundamental basis for comprehending the characteristics and actions of atoms. The phenomenon of atomic spectra is a consequence of the interaction between light and atoms, which gives rise to distinct patterns of wavelengths that are either emitted or absorbed. This phenomenon functions as a distinctive "fingerprint" for every element, enabling scientists to ascertain and measure the presence of elements in different materials.

Subsequently, it explored the various classifications of atomic spectra, encompassing absorption, emission, and scattering spectra. Measuring the wavelengths of light absorbed by a sample, known as absorption spectra, provides valuable insights into the sample's composition. In contrast, emission spectra primarily concern the wavelengths of light emitted by atoms upon their transition back to the ground state after excitation. In contrast, scattering spectra investigate the phenomenon of light deflection or scattering caused by particles or molecules, offering valuable insights into the material's structure.

Another important aspect discussed in this chapter pertains to comprehending atom emission processes. When atoms undergo excitation, the movement of their electrons between different energy levels occurs, leading to the emission of photons. The photons that are emitted carry significant information regarding the energy levels and electronic configurations of atoms. This enables researchers to investigate atomic behavior and determine the presence of elements in samples using methods such as AES.

The investigation involved the examination of energy level diagrams and transition notations to represent the intricate electron transitions occurring within atoms accurately. The diagrams presented herein visually depict energy states, electronic orbits, and permissible transitions between them. Transition notations are crucial for succinctly describing transitions and for comprehending atomic behavior and spectral lines.

Subsequently, it explored the Bohr model of the atom, a paradigm-shifting concept that significantly advanced the comprehension of atomic composition. Despite its groundbreaking nature, the Bohr model exhibited certain limitations, particularly its inadequate ability to provide accurate explanations for the behavior of multi-electron atoms. Notwithstanding these constraints, the Bohr model established the foundation for subsequent quantum mechanical models, facilitating a more profound understanding of atomic phenomena.

The chapter thoroughly discussed the quantum mechanical model and its correlation with atomic spectra. Quantum mechanics, a fundamental principle in contemporary physics, offers a comprehensive detail of atomic phenomena, encompassing electron wave functions, energy levels, and spectral lines. The model above effectively overcomes the constraints of the Bohr model and continues to serve as a fundamental instrument for comprehending the enigmas of the quantum field and its associations with atomic spectra.

Spectroscopic techniques have emerged as potent tools in investigating atomic spectra and emissions. A wide range of spectroscopic techniques, including absorption and emission spectroscopy, X-ray spectroscopy, and NMR spectroscopy, provide versatile approaches for investigating the properties of matter and its interactions with electromagnetic radiation. The field of atomic spectroscopy is primarily concerned with investigating atoms' absorption or emission of electromagnetic radiation. This area of study plays a crucial role in the identification of elements and the quantitative analysis of substances.

In summary, this chapter has thoroughly explored atomic spectra and emissions. Investigating light-matter interactions has provided significant insights into atomic structure, elemental identification, and matter dynamics. The diverse range of spectroscopic techniques demonstrated the multifunctionality and significance of these methods across various scientific disciplines. As the comprehension of atomic phenomena progresses, these insights catalyze significant breakthroughs and progress in physics, chemistry, and other disciplines.

REFERENCES

1. Aguilera, J. A., & Aragón, C., (2004). Characterization of a laser-induced plasma by spatially resolved spectroscopy of neutral atom and ion emissions.: Comparison of local and spatially integrated measurements. *Spectrochimica Acta Part B: Atomic Spectroscopy*, *59*(12), 1861–1876.
2. Antequera, T., Caballero, D., Grassi, S., Uttaro, B., & Perez-Palacios, T., (2021). Evaluation of fresh meat quality by hyperspectral imaging (HSI), nuclear magnetic resonance (NMR) and magnetic resonance imaging (MRI): A review. *Meat Science*, *172*, 108340.
3. Aragón, C., & Aguilera, J. A., (2008). Characterization of laser induced plasmas by optical emission spectroscopy: A review of experiments and methods. *Spectrochimica Acta Part B: Atomic Spectroscopy*, *63*(9), 893–916.
4. Askerka, M., Brudvig, G. W., & Batista, V. S., (2017). The O_2-evolving complex of photosystem II: Recent insights from quantum mechanics/molecular mechanics (QM/MM), extended X-ray absorption fine structure (EXAFS), and femtosecond X-ray crystallography data. *Accounts of Chemical Research*, *50*(1), 41–48.
5. Axner, O., Gustafsson, J., Omenetto, N., & Winefordner, J. D., (2004). Line strengths, A-factors, and absorption cross-sections for fine structure lines in multiplets and hyperfine structure components in lines in atomic spectrometry – a user's guide. *Spectrochimica Acta Part B: Atomic Spectroscopy*, *59*(1), 1–39.
6. Ballentine, L. E., (2014). *Quantum Mechanics: A Modern Development*. World Scientific Publishing Company, 1, 1-17.
7. Barnett, W. B., Fassel, V. A., & Kniseley, R. N., (1968). Theoretical principles of internal standardization in analytical emission spectroscopy. *Spectrochimica Acta Part B: Atomic Spectroscopy*, *23*(10), 643–664.
8. Bauche-Arnoult, C., Bauche, J., & Klapisch, M., (1982). Variance of the distributions of energy levels and of the transition arrays in atomic spectra. II. Configurations with more than two open subshells. *Physical Review A*, *25*(5), 2641.
9. Bendicho, C., & de Loos-Vollebregt, M. T., (1991). Solid sampling in electrothermal atomic absorption spectrometry using commercial atomizers. A review. *Journal of Analytical Atomic Spectrometry*, *6*(5), 353–374.

10. Berdyugina, S. V., & Solanki, S. K., (2002). The molecular Zeeman effect and diagnostics of solar and stellar magnetic fields-I. Theoretical spectral patterns in the Zeeman regime. *Astronomy & Astrophysics, 385*(2), 701–715.
11. Bersier, P. M., Howell, J., & Bruntlett, C., (1994). Tutorial review. Advanced electroanalytical techniques versus atomic absorption spectrometry, inductively coupled plasma atomic emission spectrometry and inductively coupled plasma mass spectrometry in environmental analysis. *Analyst, 119*(2), 219–232.
12. Bigio, I. J., & Mourant, J. R., (1997). Ultraviolet and visible spectroscopies for tissue diagnostics: Fluorescence spectroscopy and elastic-scattering spectroscopy. *Physics in Medicine and Biology, 42*(5), 803.
13. Bings, N. H., Bogaerts, A., & Broekaert, J. A., (2010). Atomic spectroscopy: A review. *Analytical Chemistry, 82*(12), 4653–4681.
14. Casasayas-Barris, N., Pallé, E., Yan, F., Chen, G., Kohl, S., Stangret, M., & Osorio, M. Z., (2019). Atmospheric characterization of the ultra-hot Jupiter MASCARA-2b/KELT-20b-detection of CaII, FeII, NaI, and the Balmer series of H (Hα, Hβ, and Hγ) with high-dispersion transit spectroscopy. *Astronomy & Astrophysics, 628*, A9.
15. Chakrabarti, C. L., Gilmutdinov, A. K., & Hutton, J. C., (1993). Digital imaging of atomization processes in electrothermal atomizers for atomic absorption spectrometry. *Analytical Chemistry, 65*(6), 716–723.
16. Chinni, R. C., (2012). A simple LIBS (laser-induced breakdown spectroscopy) laboratory experiment to introduce undergraduates to calibration functions and atomic spectroscopy. *Journal of Chemical Education, 89*(5), 678–680.
17. Christian, G. D., (1969). Medicine, trace elements, and atomic absorption spectroscopy. *Analytical Chemistry, 41*(1), 24A–40A.
18. Cowan, R. D., (1968). Theoretical calculation of atomic spectra using digital computers. *JOSA, 58*(6), 808–818.
19. Cruzeiro, V. W. D., Wildman, A., Li, X., & Paesani, F., (2021). Relationship between hydrogen-bonding motifs and the 1b1 splitting in the X-ray emission spectrum of liquid water. *The Journal of Physical Chemistry Letters, 12*(16), 3996–4002.
20. David, D. J., (1958). Determination of zinc and other elements in plants by atomic-absorption spectroscopy. *Analyst, 83*(993), 655–661.

21. Epstein, P. S., (1926). The Stark effect from the point of view of Schroedinger's quantum theory. *Physical Review*, *28*(4), 695.
22. Eremin, M. V., Iglamov, V. V., & Fayzullin, M. A., (2011). Energy level diagram, wavefunctions, and probabilities of optical transitions in KCuF 3. *Optics and Spectroscopy*, *111*, 393–396.
23. Ezawa, M., (2012). Spin-valley optical selection rule and strong circular dichroism in silicene. *Physical Review B*, *86*(16), 161407.
24. Falk, H., (1991). Tandem sources using electrothermal atomizers: analytical capabilities and limitations. *Journal of Analytical Atomic Spectrometry*, *6*(8), 631–635.
25. Fried, S. D., & Boxer, S. G., (2015). Measuring electric fields and noncovalent interactions using the vibrational Stark effect. *Accounts of Chemical Research*, *48*(4), 998–1006.
26. Friedman, A., Gover, A., Kurizki, G., Ruschin, S., & Yariv, A., (1988). Spontaneous and stimulated emission from quasifree electrons. *Reviews of Modern Physics*, *60*(2), 471.
27. Fujihara, J., & Nishimoto, N., (2020). Total antimony analysis by hydride generation-microwave plasma-atomic emission spectroscopy with applications. *Microchemical Journal*, *157*, 104992.
28. Fujisawa, T., Austing, D. G., Tokura, Y., Hirayama, Y., & Tarucha, S., (2002). Allowed and forbidden transitions in artificial hydrogen and helium atoms. *Nature*, *419*(6904), 278–281.
29. Garcia, J. D., & Mack, J. E., (1965). Energy level and line tables for one-electron atomic spectra. *JOSA*, *55*(6), 654–685.
30. Gerothanassis, I. P., Troganis, A., Exarchou, V., & Barbarossou, K., (2002). Nuclear magnetic resonance (NMR) spectroscopy: Basic principles and phenomena, and their applications to chemistry, biology, and medicine. *Chemistry Education Research and Practice*, *3*(2), 229–252.
31. Gilmutdinov, A. K., Mrasov, R. M., Somov, A. R., Chakrabarti, C. L., & Hutton, J. C., (1995). Three-dimensional modelling of the analyte dynamics in electrothermal atomizers for analytical spectrometry: influence of physical factors. *Spectrochimica Acta Part B: Atomic Spectroscopy*, *50*(13), 1637–1654.
32. Ginocchio, J. N., & Kirson, M. W., (1980). An intrinsic state for the interacting boson model and its relationship to the Bohr-Mottelson model. *Nuclear Physics A*, *350*(1–2), 31–60.

33. Govindaraju, V., Young, K., & Maudsley, A. A., (2000). Proton NMR chemical shifts and coupling constants for brain metabolites. *NMR in Biomedicine: An International Journal Devoted to the Development and Application of Magnetic Resonance In Vivo*, *13*(3), 129–153.

34. Grinshtein, I. L., Vilpan, Y. A., Saraev, A. V., & Vasilieva, L. A., (2001). Direct atomic absorption determination of cadmium and lead in strongly interfering matrices by double vaporization with a two-step electrothermal atomizer. *Spectrochimica Acta Part B: Atomic Spectroscopy*, *56*(3), 261–274.

35. Hehenberger, M., McIntosh, H. V., & Brändas, E., (1974). Weyl's theory applied to the Stark effect in the hydrogen atom. *Physical Review A*, *10*(5), 1494.

36. Hell, S. W., & Wichmann, J., (1994). Breaking the diffraction resolution limit by stimulated emission: Stimulated-emission-depletion fluorescence microscopy. *Optics Letters*, *19*(11), 780–782.

37. Hermsmeier, B., Fadley, C. S., Krause, M. O., Jimenez-Mier, J., Gerard, P., & Manson, S. T., (1988). Direct evidence from gas-phase atomic spectra for an unscreened intra-atomic origin of outer-core multiplet splittings in solid manganese compounds. *Physical Review Letters*, *61*(22), 2592.

38. Hoenig, M., & de Kersabiec, A. M., (1996). Sample preparation steps for analysis by atomic spectroscopy methods: Present status. *Spectrochimica Acta Part B: Atomic Spectroscopy*, *51*(11), 1297–1307.

39. Hong, J., Jin, C., Yuan, J., & Zhang, Z., (2017). Atomic defects in two - dimensional materials: From single - atom spectroscopy to functionalities in opto - /electronics, nanomagnetism, and catalysis. *Advanced Materials*, *29*(14), 1606434.

40. Kalfas, C. A., Axiotis, M., & Tsabaris, C., (2016). SPECTRW: A software package for nuclear and atomic spectroscopy. *Nuclear Instruments and Methods in Physics Research Section A: Accelerators, Spectrometers, Detectors, and Associated Equipment*, *830*, 265–274.

41. Kaufman, V., & Edlén, B., (1974). Reference wavelengths from atomic spectra in the range 15 Å to 25,000 Å. *Journal of Physical and Chemical Reference Data*, *3*(4), 825–895.

42. Kennicutt Jr, R. C., (1992). The integrated spectra of nearby galaxies-General properties and emission-line spectra. *The Astrophysical Journal*, *388*, 310–327.

43. Kulander, K. C., & Rescigno, T. N., (1991). Effective potentials for time-dependent calculations of multiphoton processes in atoms. *Computer Physics Communications*, *63*(1–3), 523–528.
44. Layzer, D., (1959). On a screening theory of atomic spectra. *Annals of Physics*, *8*(2), 271–296.
45. Leapman, R. D., Grunes, L. A., & Fejes, P. L., (1982). Study of the L 23 edges in the 3 d transition metals and their oxides by electron-energy-loss spectroscopy with comparisons to theory. *Physical Review B*, *26*(2), 614.
46. Lewen, N., (2011). The use of atomic spectroscopy in the pharmaceutical industry for the determination of trace elements in pharmaceuticals. *Journal of Pharmaceutical and Biomedical Analysis*, *55*(4), 653–661.
47. Lloyd, S., Babiker, M., & Yuan, J., (2012). Quantized orbital angular momentum transfer and magnetic dichroism in the interaction of electron vortices with matter. *Physical Review Letters*, *108*(7), 074802.
48. Mann, M., Hendrickson, R. C., & Pandey, A., (2001). Analysis of proteins and proteomes by mass spectrometry. *Annual Review of Biochemistry*, *70*(1), 437–473.
49. Markeev, P. A., Najafidehaghani, E., Gan, Z., Sotthewes, K., George, A., Turchanin, A., & de Jong, M. P., (2021). Energy-level alignment at interfaces between transition-metal dichalcogenide monolayers and metal electrodes studied with Kelvin probe force microscopy. *The Journal of Physical Chemistry C*, *125*(24), 13551–13559.
50. Marsman, A., Horbatsch, M., & Hessels, E. A., (2015). The effect of quantum-mechanical interference on precise measurements of the n = 2 triplet p fine structure of helium. *Journal of Physical and Chemical Reference Data*, *44*(3).
51. Maurya, A., Kesharwani, L., & Mishra, M. K., (2018). Analysis of heavy metal in soil through atomic absorption spectroscopy for forensic consideration. *Int. J. Res. Appl. Sci. Eng. Technol.*, *6*(6), 1188–1192.
52. McKagan, S. B., Perkins, K. K., & Wieman, C. E., (2008). Why we should teach the Bohr model and how to teach it effectively. *Physical Review Special Topics-Physics Education Research*, *4*(1), 010103.
53. McNally, J., & Holcombe, J. A., (1987). Existence of microdroplets and dispersed atoms on the graphite surface in electrothermal atomizers. *Analytical Chemistry*, *59*(8), 1105–1112.

54. Merrill, P. W., & Wilson, O. C., (1934). Comparison of the Paschen and the Balmer series of hydrogen lines in stellar spectra. *Astrophysical Journal, vol. 80, p. 19, 80*, 19.

55. Meyer, B., & Peters, T., (2003). NMR spectroscopy techniques for screening and identifying ligand binding to protein receptors. *Angewandte Chemie International Edition, 42*(8), 864–890.

56. Middleton, W. D., & Price, D. T., (1996). Identification of activity areas by multi-element characterization of sediments from modern and archaeological house floors using inductively coupled plasma-atomic emission spectroscopy. *Journal of Archaeological Science, 23*(5), 673–687.

57. Newman, S., Stolper, E. M., & Epstein, S., (1986). Measurement of water in rhyolitic glasses; calibration of an infrared spectroscopic technique. *American Mineralogist, 71*(11–12), 1527–1541.

58. Nilsson, M., & Kröll, S., (2005). Solid state quantum memory using complete absorption and re-emission of photons by tailored and externally controlled inhomogeneous absorption profiles. *Optics Communications, 247*(4–6), 393–403.

59. Noda, S., Fujita, M., & Asano, T., (2007). Spontaneous-emission control by photonic crystals and nanocavities. *Nature Photonics, 1*(8), 449–458.

60. Onorato, P., Malgieri, M., & De Ambrosis, A., (2015). Measuring the hydrogen Balmer series and Rydberg's constant with a homemade spectrophotometer. *European Journal of Physics, 36*(5), 058001.

61. Ozbek, N., & Akman, S., (2012). Method development for the determination of fluorine in water samples via the molecular absorption of strontium monofluoride formed in an electrothermal atomizer. *Spectrochimica Acta Part B: Atomic Spectroscopy, 69*, 32–37.

62. Parigger, C. G., (2013). Atomic and molecular emissions in laser-induced breakdown spectroscopy. *Spectrochimica Acta Part B: Atomic Spectroscopy, 79*, 4–16.

63. Parsons, M. L., McCarthy, W. J., & Winefordner, J. D., (1966). Approximate half-intensity widths of a number of atomic spectral lines used in atomic-emission and atomic-absorption flame spectrometry. *Applied Spectroscopy, 20*(4), 223–230.

64. Penache, C., Miclea, M., Bräuning-Demian, A., Hohn, O., Schössler, S., Jahnke, T., & Schmidt-Böcking, H., (2002). Characterization of a high-pressure microdischarge using diode laser atomic absorption spectroscopy. *Plasma Sources Science and Technology, 11*(4), 476.

65. Peterson, J. R., & Fabian, A. C., (2006). X-ray spectroscopy of cooling clusters. *Physics Reports*, *427*(1), 1–39.
66. Porra, R. J., Thompson, W. A. A., & Kriedemann, P. E., (1989). Determination of accurate extinction coefficients and simultaneous equations for assaying chlorophylls a and b extracted with four different solvents: verification of the concentration of chlorophyll standards by atomic absorption spectroscopy. *Biochimica et Biophysica Acta (BBA)-Bioenergetics*, *975*(3), 384–394.
67. Rezac, J., & Hobza, P., (2012). Advanced corrections of hydrogen bonding and dispersion for semiempirical quantum mechanical methods. *Journal of Chemical Theory and Computation*, *8*(1), 141–151.
68. Rosenzweig, N., & Porter, C. E., (1960). "Repulsion of energy levels" in complex atomic spectra. *Physical Review*, *120*(5), 1698.
69. Rudzikas, Z., (2007). Theoretical atomic spectroscopy. *Theoretical Atomic Spectroscopy*.
70. Ruhaak, L. R., Xu, G., Li, Q., Goonatilleke, E., & Lebrilla, C. B., (2018). Mass spectrometry approaches to glycomic and glycoproteomic analyses. *Chemical Reviews*, *118*(17), 7886–7930.
71. Savall-Alemany, F., Domènech-Blanco, J. L., Guisasola, J., & Martínez-Torregrosa, J., (2016). Identifying student and teacher difficulties in interpreting atomic spectra using a quantum model of emission and absorption of radiation. *Physical Review Physics Education Research*, *12*(1), 010132.
72. Schutte, C. J. H., Bertie, J. E., Bunker, P. R., Hougen, J. T., Mills, I. M., Watson, J. K. G., & Winnewisser, B. P., (1997). Notations and conventions in molecular spectroscopy: Part 1. General spectroscopic notation (IUPAC recommendations 1997). *Pure and Applied Chemistry*, *69*(8), 1633–1640.
73. Shkirskiy, V., King, A. D., Gharbi, O., Volovitch, P., Scully, J. R., Ogle, K., & Birbilis, N., (2015). Revisiting the electrochemical impedance spectroscopy of magnesium with online inductively coupled plasma atomic emission spectroscopy. *ChemPhysChem*, *16*(3), 536–539.
74. Sobel'Man, I. I., (2016). *Introduction to the Theory of Atomic Spectra: International Series of Monographs in Natural Philosophy* (Vol. 40) pp. 1-17.

75. Solovyev, D., Anikin, A., Zalialiutdinov, T., & Labzowsky, L., (2020). Importance of nonresonant corrections for the description of atomic spectra. *Journal of Physics B: Atomic, Molecular, and Optical Physics*, *53*(12), 125002.

76. Stuart, B., (2000). Infrared spectroscopy. *Kirk - Othmer Encyclopedia of Chemical Technology*.

77. Suenaga, K., & Koshino, M., (2010). Atom-by-atom spectroscopy at graphene edge. *Nature*, *468*(7327), 1088–1090.

78. Tittarelli, P., Biffi, C., & Kmetov, V., (1994). Vaporization of silicon and germanium as molecular species in electrothermal atomizers. *Journal of Analytical Atomic Spectrometry*, *9*(3), 443–449.

79. Uden, P. C., (1992). *Element-Specific Chromatographic Detection by Atomic Emission Spectroscopy* (No. CONF-900402). Washington, DC (United States); American Chemical Society.

80. Vil'pan, Y. A., Grinshtein, I. L., Akatov, A. A., & Gucer, S., (2005). Direct atomic absorption determination of mercury in drinking water and urine using a two-step electrothermal atomizer. *Journal of Analytical Chemistry*, *60*, 38–44.

81. Wang, K., Sun, D. W., Pu, H., & Wei, Q., (2017). Principles and applications of spectroscopic techniques for evaluating food protein conformational changes: A review. *Trends in Food Science & Technology*, *67*, 207–219.

82. Wende, H., (2004). Recent advances in X-ray absorption spectroscopy. *Reports on Progress in Physics*, *67*(12), 2105.

83. Wienke, D., Lucasius, C., & Kateman, G., (1992). Multicriteria target vector optimization of analytical procedures using a genetic algorithm: Part I. Theory, numerical simulations and application to atomic emission spectroscopy. *Analytica Chimica Acta*, *265*(2), 211–225.

84. Wigner, E., (2012). *Group Theory and its Application to the Quantum Mechanics of Atomic Spectra* (Vol. 5), pp. 1-10.

85. Yngström, S., (1994). Theory of atomic spectral emission intensity. *International Journal of Theoretical Physics*, *33*, 1479–1501.

86. Yuzefovsky, A. I., Lonardo, R. F., Wang, M., & Michel, R. G., (1994). Determination of ultra-trace amounts of cobalt in ocean water by laser-excited atomic fluorescence spectrometry in a graphite electrothermal atomizer with semi on-line flow injection preconcentration. *Journal of Analytical Atomic Spectrometry*, *9*(11), 1195–1202.

87. Zhu, S. Y., Narducci, L. M., & Scully, M. O., (1995). Quantum-mechanical interference effects in the spontaneous-emission spectrum of a driven atom. *Physical Review A, 52*(6), 4791.

CHAPTER 4

Intermolecular Forces

Contents

In the intricate world of chemistry, the interactions between molecules play a fundamental role in determining the physical and chemical properties of substances. These interactions are governed by a set of forces known as intermolecular forces. While the concept of chemical bonds holds atoms together within molecules, intermolecular forces are responsible for the attractions and repulsions between different molecules. These forces are subtle yet crucial, influencing a wide range of phenomena, from the behavior of gases and liquids to the structure and properties of complex biomolecules. In this chapter, we will delve into the various types, their significance, and their impact on the behavior of matter at the molecular level. Understanding these forces is not only essential for explaining the physical world around us but also for advancing fields such as materials science, pharmaceuticals, and environmental chemistry (Lynden-Bell, 1977) (Figure 4.1).

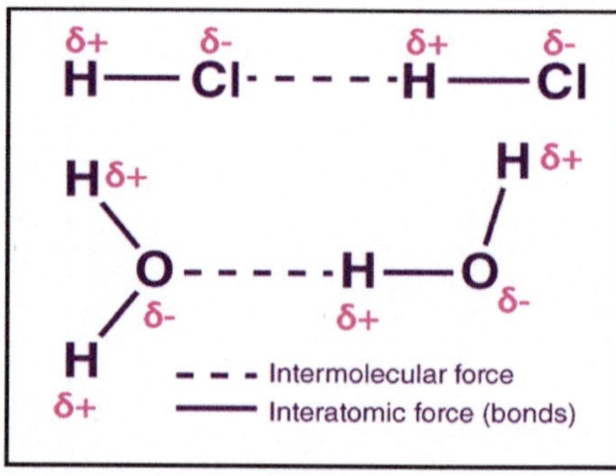

Figure 4.1. *Intermolecular force.*

Source: Boris, Creative Commons License.

Intermolecular forces are separate from chemical bonds, which maintain the cohesion between atoms within molecules. Conversely, these forces originate from the molecular interactions themselves. The magnitude and characteristics of intermolecular forces are contingent upon the molecular composition and the geometrical and dimensional attributes of the molecules (Akinmade & Nicholson, 1993).

4.1. VAN DER WAALS FORCES

Although the least strong among intermolecular forces, Van der Waals forces are widely observed in numerous substances. Two fundamental categories of Van der Waals forces exist: London dispersion forces and dipole-dipole interactions, the two types of intermolecular forces that play significant roles in molecular interactions. London dispersion forces, also known as van der Waals forces, arise from temporary fluctuations in electron distribution within molecules, forming temporary dipoles (Berland et al., 2015).

4.1.1. London Dispersion Forces

London dispersion forces are universally observed in all molecules, irrespective of their polarity. These forces result from transient variations in electron distribution within molecules, forming transient dipoles. The temporary dipoles in molecules result in corresponding dipoles in adjacent molecules, thereby giving rise to attractive intermolecular forces. The magnitude of London dispersion forces is positively correlated with the dimensions and configuration of molecules, resulting in enhanced intermolecular interactions for larger and more elongated molecular structures (Von Lilienfeld et al., 2004) (Figure 4.2).

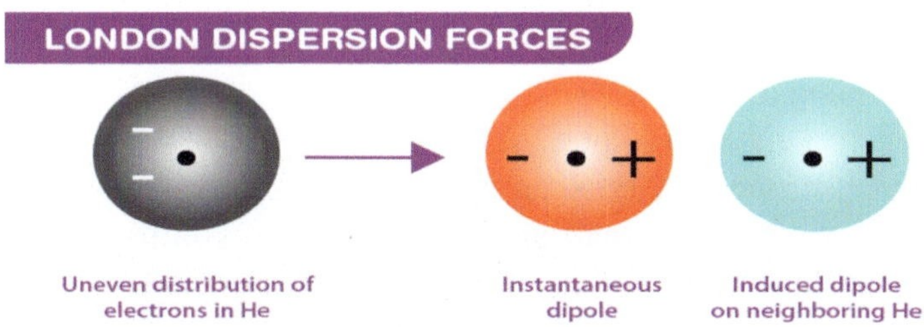

Figure 4.2. *Illustration of London dispersion forces.*

Source: BYJU'S, Creative Commons License.

4.1.2. Dipole-Dipole Interactions

Dipole-dipole interactions manifest in polar molecules, wherein one extremity of the molecule exhibits a partial positive charge, while the opposite extremity exhibits a partial negative charge.

The attractive interaction between polar molecules occurs when the positive pole of one molecule is drawn towards the negative pole of another molecule, leading to a more robust force than the London dispersion forces. Dipole-dipole interactions play a substantial role in shaping polar substances' characteristics and influence their behavior in diverse environments (Browaeys et al., 2016) (Figure 4.3).

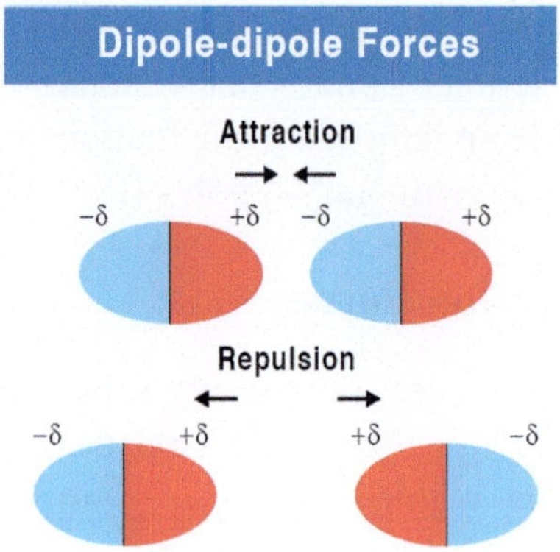

Figure 4.3. *Illustration of dipole-dipole interaction.*

Source: Chemistry Learner, Creative Commons License.

4.2. HYDROGEN BONDING

Hydrogen bonding is a distinctive form of dipole-dipole interaction that manifests when a hydrogen atom is covalently bonded to an atom with notably high electronegativity, such as nitrogen, oxygen, or fluorine. The hydrogen atom, possessing a partial positive charge, exhibits a pronounced affinity towards the lone pair of electrons on the electronegative atom within a separate molecule (Figure 4.4).

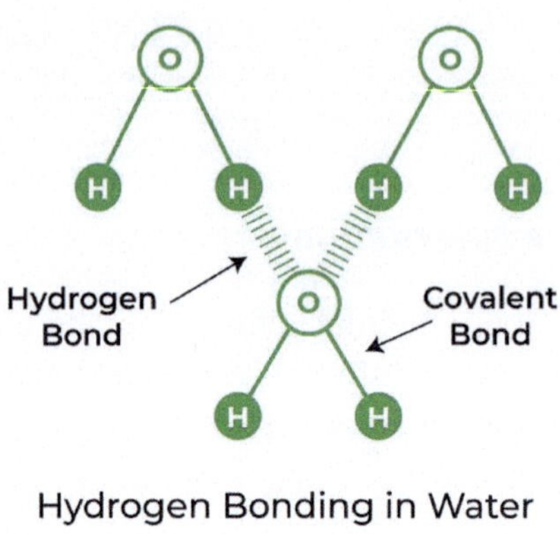

Figure 4.4. *Schematic representation of hydrogen bonding.*

Source: GeeksforGeeks, Creative Commons License.

Hydrogen bonding is recognized for its significantly greater strength than conventional dipole-dipole interactions, resulting in distinctive characteristics observed in substances that manifest this particular form of bonding.

It plays a pivotal role in numerous fundamental biological phenomena, including the maintenance of DNA's structure and stability, the intricate process of protein folding, and the cohesive characteristics exhibited by water. The role of hydrogen bonding in the behavior of substances, such as alcohol and ammonia, is of utmost importance, as it significantly impacts their physical properties (Kuhn, 1952).

4.3. ION-DIPOLE INTERACTIONS

Ion-dipole interactions manifest as attractive forces between charged particles, specifically ions and polar molecules.

In a solution comprising ionic compounds, the cations and anions engage in electrostatic interactions with the partial charges on the polar molecules, thereby facilitating the dissolution of the ionic compound. Ion-dipole interactions play a pivotal role in phenomena such as solvation, wherein ions dissolve in a solvent, as well as in diverse industrial processes that involve electrolytes and polar solvents (Krygowski & Fawcett, 1975) (Figure 4.5).

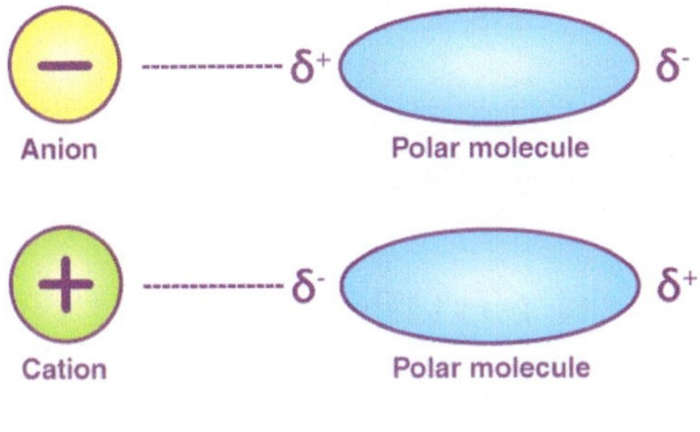

Figure 4.5. *Illustration of ion-dipole interactions.*

Source: Byjus, Creative Commons License.

4.4. SURFACE TENSION AND CAPILLARY ACTION

Surface tension, a characteristic of liquids wherein they develop a cohesive "skin" or "film" on their surface, is regulated by intermolecular forces. The cohesive force observed in liquids arises from the intermolecular interactions occurring at the surface of the liquid, resulting in resistance to penetration and deformation (Figure 4.6).

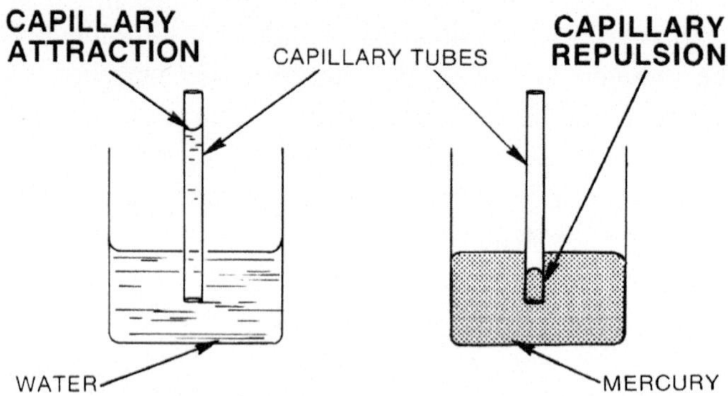

Figure 4.6. *Schematic representation of capillary action.*

Source: Chemistry Libre Texts, Creative Commons License.

Capillary action is a phenomenon that is influenced by intermolecular forces. This phenomenon pertains to the capacity of liquids to exhibit upward flow in confined spaces, such as within a slender conduit, in opposition to the force of gravity. Capillary action is a phenomenon that occurs as a result of the equilibrium between adhesive forces, which exist between the liquid and the tube, and cohesive forces, which are present within the liquid. The magnitude of these forces dictates the vertical distance the liquid will ascend within the tube (Mondal et al., 2009).

4.5. EXAMPLES OF COMPOUNDS INVOLVING MULTIPLE FORCES

Several compounds exhibit multiple intermolecular forces due to their complex molecular structures. Some examples are given in subsections.

4.5.1. Water (H_2O)

Water is a classic example of a compound that displays multiple intermolecular forces. It exhibits hydrogen bonding between water molecules due to the strong electronegativity of oxygen and the presence of hydrogen atoms bonded to it. Additionally, water molecules also experience

London dispersion forces and dipole-dipole interactions because they are polar molecules (O'Connell & Prausnitz, 1969) (Figure 4.7).

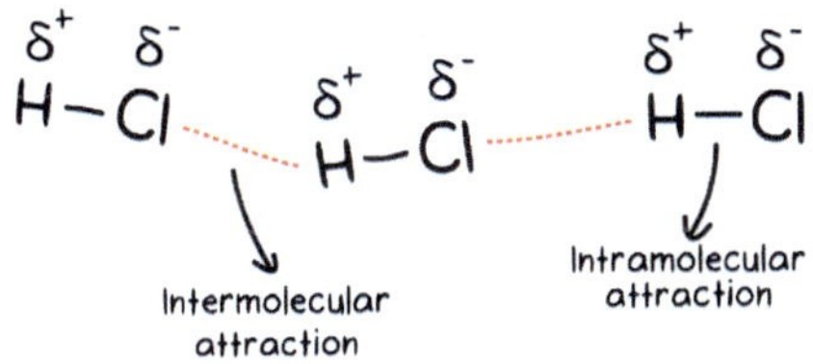

Figure 4.7. *Illustration of intermolecular forces in water molecules.*

Source: Khan Academy, Creative Commons License.

4.5.2. Ammonia (NH_3)

Ammonia is another molecule that showcases various intermolecular forces. It undergoes hydrogen bonding between ammonia molecules because of the hydrogen atoms bonded to nitrogen. Like water, ammonia also has London dispersion forces and dipole-dipole interactions due to its polarity (Ikuta, 1987) (Figure 4.8).

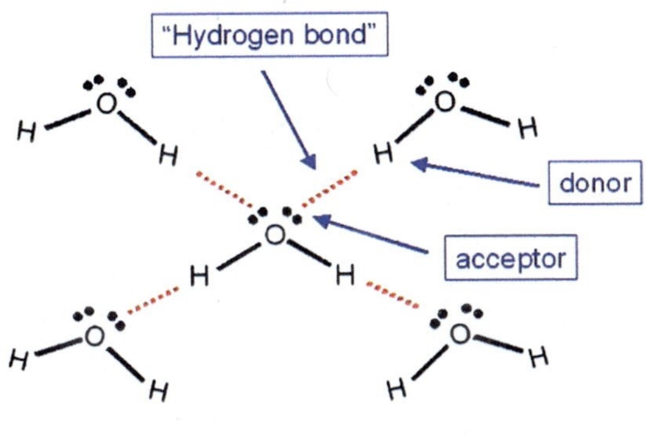

Figure 4.8. *Ammonia molecules showing multiple forces.*

Source: Socratic, Creative Commons License.

4.5.3. Hydrogen Chloride (HCl)

Hydrogen chloride is a polar molecule with a partial positive charge on hydrogen and a partial negative charge on chlorine. This polarity leads to dipole-dipole interactions between HCl molecules. Additionally, London dispersion forces also come into play (Hutson & Howard, 1982) (Figure 4.9).

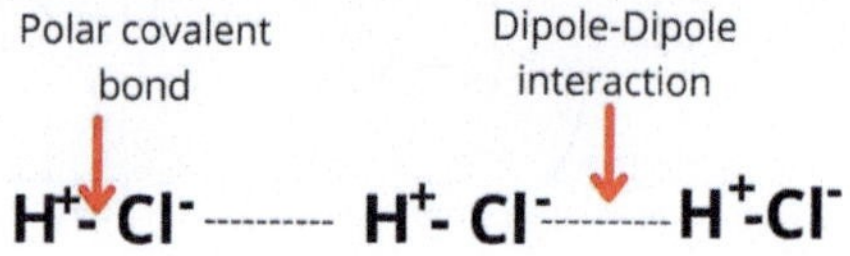

Figure 4.9. *Illustration of HCL structure.*

Source: Techie Scientist, Creative Commons License.

4.5.4. Methane (CH_4)

Methane is a nonpolar molecule composed of carbon and hydrogen atoms. While it lacks permanent dipoles, it still exhibits London dispersion forces due to the temporary fluctuations in electron distribution around its atoms. Although weaker than dipole-dipole interactions, these dispersion forces are present (Figure 4.10).

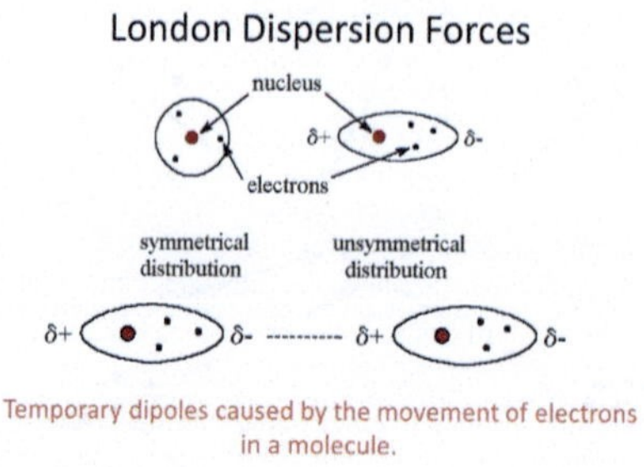

Figure 4.10. *Representation of intermolecular forces in methane.*

Source: Techie Scientist, Creative Commons License.

4.5.5. Carbon Tetrachloride (CCl_4)

Carbon tetrachloride is a nonpolar molecule similar to methane. It experiences London dispersion forces because of its electron cloud fluctuations despite having no permanent dipole moment.

4.5.6. Acetic Acid (CH_3COOH)

Acetic acid is a polar molecule containing both hydrogen bonding and dipole-dipole interactions. The hydrogen atom in the –OH group can engage in hydrogen bonding, while the carbonyl group (C=O) contributes to dipole-dipole interactions. Furthermore, London dispersion forces are also present.

4.5.7. Ethanol (C_2H_5OH)

Ethanol displays a combination of hydrogen bonding and dipole-dipole interactions. The –OH group in ethanol can participate in hydrogen bonding, while the molecule's overall polarity leads to dipole-dipole forces. Like most organic compounds, ethanol also has London dispersion forces (Kamogawa et al., 1987) (Figure 4.11).

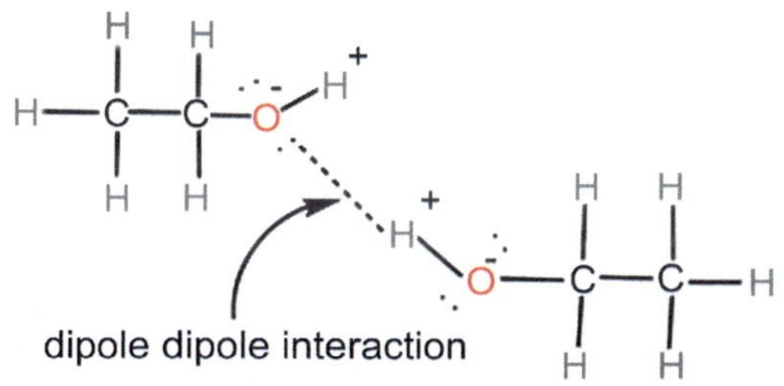

Figure 4.11. *Illustration of intermolecular forces in ethanol.*

Source: Chem White, Creative Commons License.

These examples illustrate that many compounds can exhibit multiple intermolecular forces simultaneously, depending on their molecular structures and functional groups. The presence and relative strengths of these forces influence the physical properties and behavior of these compounds.

4.6. SUMMARY

The chapter on intermolecular forces provides a comprehensive exploration of the various types of interactions that occur between molecules. It delves into the distinctions between these forces and chemical bonds, emphasizing how intermolecular forces govern attractions and repulsions between different molecules. The chapter goes on to detail the specific types of intermolecular forces, such as Van der Waals forces, which include London dispersion forces and dipole-dipole interactions. It explains how these forces are influenced by molecular composition, geometry, and size. Hydrogen bonding, a unique and powerful form of dipole-dipole interaction, is discussed extensively, highlighting its significance in biology and the behavior of substances like water, alcohol, and ammonia. The chapter also explores ion-dipole interactions, surface tension, and capillary action, elucidating how these forces impact solvation, industrial processes, and the behavior of liquids in confined spaces. Furthermore, the chapter provides examples of compounds that exhibit multiple intermolecular forces, showcasing the practical relevance of this knowledge in various fields of science and technology. Overall, the chapter on intermolecular forces serves as a fundamental guide to understanding the subtle yet crucial forces that shape the behavior of matter at the molecular level, with implications for fields ranging from chemistry to materials science and beyond.

REFERENCES

1. Akinmade, A. O., & Nicholson, J. W., (1993). Glass-ionomer cements as adhesives: Part I Fundamental aspects and their clinical relevance. *Journal of Materials Science: Materials in Medicine, 4*(2), 95–101.
2. Berland, K., Cooper, V. R., Lee, K., Schröder, E., Thonhauser, T., Hyldgaard, P., & Lundqvist, B. I., (2015). van der Waals forces in density functional theory: A review of the vdW-DF method. *Reports on Progress in Physics, 78*(6), 066501.
3. Browaeys, A., Barredo, D., & Lahaye, T., (2016). Experimental investigations of dipole–dipole interactions between a few Rydberg atoms. *Journal of Physics B: Atomic, Molecular, and Optical Physics, 49*(15), 152001.
4. Hutson, J. M., & Howard, B. J., (1982). Anisotropic intermolecular forces: I. Rare gas-hydrogen chloride systems. *Molecular Physics, 45*(4), 769–790.
5. Ikuta, S., (1987). Electron correlation effect on the geometries and energetics: Proton - bound ammonia dimer $(H_3N–H–NH_3)^+$. *The Journal of Chemical Physics, 87*(3), 1900–1901.
6. Krygowski, T. M., & Fawcett, W. R., (1975). Complementary Lewis acid-base description of solvent effects. I. Ion-ion and ion-dipole interactions. *Journal of the American Chemical Society, 97*(8), 2143–2148.
7. Kuhn, L. P., (1952). The hydrogen bond: I. Intra-and Intermolecular hydrogen bonds in alcohols. *Journal of the American Chemical Society, 74*(10), 2492–2499.
8. Lynden-Bell, R. M., (1977). Vibrational relaxation and line widths in liquids dephasing by intermolecular forces. *Molecular Physics, 33*(4), 907–921.
9. Mondal, S. K., Mitra, A., Singh, N., Sarkar, S. N., & Kapur, P., (2009). Optical fiber nanoprobe preparation for near-field optical microscopy by chemical etching under surface tension and capillary action. *Optics Express, 17*(22), 19470–19475.
10. Morrow, N. R., (1970). Physics and thermodynamics of capillary action in porous media. *Industrial & Engineering Chemistry, 62*(6), 32–56.
11. O'Connell, J. P., & Prausnitz, J. M., (1969). Intermolecular forces in water vapor. *Industrial & Engineering Chemistry Fundamentals, 8*(3), 453–460.

12. Von Lilienfeld, O. A., Tavernelli, I., Rothlisberger, U., & Sebastiani, D., (2004). Optimization of effective atom centered potentials for London dispersion forces in density functional theory. *Physical Review Letters, 93*(15), 153004.

CHAPTER 5

Chemical Bonding

Contents

5.1. INTRODUCTION

Chemical bonding is essential to understanding the behavior and interactions of atoms, molecules, and compounds. The cohesive force that forms chemical compounds includes diatomic molecules, complex biomolecules, and even materials. Chemical bonding analysis helps comprehend substances' characteristics and reactivity, advancing science and industry.

Electrons in atoms determine the dynamics of chemical bonding. Electrons orbit the positively charged nucleus of an atom. A fully populated outer electron shell lowers energy levels and stabilizes atoms. Chemical bonds arise when atoms use diverse bonding methods to stabilize (Levy Nahum et al., 2010). Chemical bonding can be ionic, covalent, or metallic. Ionic bonding involves the transfer of electrons between atoms to generate ions with opposite charges. Electrostatic forces then bind these ions. Covalent bonding, in which atoms share electrons, creates molecular structures with strong and precise directional bonds. Electrons travel freely in metal lattice structures during metallic bonding. Electron delocalization helps form a coherent lattice structure where atoms share electrons.

Numerous historical and theoretical advances have affected chemical bonding understanding. The Lewis electron-dot structure and the octet rule were key hypotheses that explained simple compound bonding. Scientific understanding has led to increasingly complicated theoretical frameworks like molecular orbital and valence bond theories. Several models explain complex molecules and solids' bonding phenomena. Chemical bonding affects several academic fields beyond chemistry. Materials science's bonding mechanisms determine material attributes like metals' mechanical strength and electrical conductivity or ceramics' thermal insulation. Chemical bonding is fundamental to the structure and function of biological macromolecules like proteins and DNA in biochemistry.

Tip

The word "bond" itself comes from the Old English word "Bendan," meaning "to tie" or "to bind." This reflects the idea that chemical bonds hold atoms together in a cohesive manner.

Quantum mechanics and computational chemistry advancements are promoting chemical bonding. Knowledge increases the ability to generate new materials, medications, and life's complicated mechanisms (Özmen, 2004).

5.2. LEWIS DOT SYMBOLS

The periodic table and electronic configuration concepts have helped chemists understand molecule and compound creation. The famous American chemist Gilbert Lewis postulated that atoms unite to stabilize their electron configurations. Isoelectronicity is the most stable state where an atom has as many electrons as a noble gas. During chemical bonding, the outermost electrons of atoms engage. (Shultz & Gere, 2015).

Lewis dot symbols, invented by Lewis, are used by chemists to track valence electrons in chemical reactions and keep electronic count. The Lewis dot sign shows an atom's elemental symbol and valence electrons as single dots. Figure 5.1 shows Lewis dot symbols for representative elements and noble gases. Besides helium, each atom's valence electron count equals its group number. Li, a Group 1A element, has one dot for its valence electron. From Group 2A of the periodic table, Beryllium has a valence electron configuration with two electrons, represented by two dots. The remaining elements in each group follow this pattern. Similar outer electron configurations result in similar Lewis dot symbols. Transition metals, lanthanides, and actinides have partial inner electron shells. Thus, Lewis dot symbols are rarely suitable for depicting these aspects (Pearce et al., 1996).

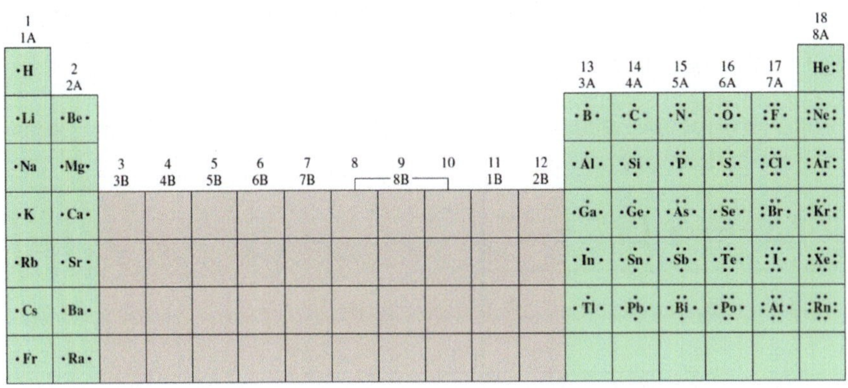

Figure 5.1. *Lewis dot symbols for the representative elements and the noble gases. The number of unpaired dots corresponds to the number of bonds an atom of the element can form in a compound.*

Source: Harald, Creative Commons License.

5.3. THE IONIC BOND

Elements with low ionization energies tend to form cations, whereas elements with high electron affinities tend to form anions. Typically, alkali, and alkaline earth metals tend to undergo cation formation in ionic compounds, while halogens and oxygen are commonly observed to undergo anion formation. As a result, the combination of a metal from Group 1A or Group 2A with either a halogen or oxygen creates a diverse range of ionic compounds. The term "ionic bond" refers to the electrostatic attraction that enables the cohesion of ions within a compound of an ionic character. A notable illustration can be witnessed in the chemical reaction involving lithium and fluorine, resulting in the synthesis of lithium fluoride. The compound mentioned above, characterized as a hazardous white powder, is employed to decrease the melting points of solder and manufacture ceramics. The electronic configuration of lithium is commonly expressed as $1s^2\ 2s^1$, while the electron configuration of fluorine is typically represented as $1s^2\ 2s^2\ 2p^5$. The empirical findings suggest that upon the interaction between lithium and fluorine atoms, the valence electron located in the outer 2s orbital of lithium undergoes a process of electron transfer to the fluorine atom. The reaction is illustrated using Lewis dot symbols (Coll & Treagust, 2003).

$$\begin{array}{ccccc} \cdot\mathrm{Li} & + & :\ddot{\underset{\cdot\cdot}{\mathrm{F}}}\cdot & \longrightarrow & \mathrm{Li}^{+}\ :\ddot{\underset{\cdot\cdot}{\mathrm{F}}}:^{-} \quad \text{(or LiF)} \\ 1s^2 2s^1 & & 1s^2 2s^2 2p^5 & & 1s^2 \quad 1s^2 2s^2 2p^6 \end{array} \tag{4.1}$$

For convenience, imagine that this reaction occurs in separate steps—fist, the ionization of Li:

$$\boxed{\cdot\mathrm{Li} \longrightarrow \mathrm{Li}^{+} + e^{-}}$$

And then the acceptance of an electron by F:

$$:\ddot{\underset{\cdot\cdot}{\mathrm{F}}}\cdot + e^{-} \longrightarrow :\ddot{\underset{\cdot\cdot}{\mathrm{F}}}:^{-}$$

Next, imagine the two separate ions joining to form a LiF unit:

$$\mathrm{Li}^{+} + :\ddot{\underset{\cdot\cdot}{\mathrm{F}}}:^{-} \longrightarrow \mathrm{Li}^{+}:\ddot{\underset{\cdot\cdot}{\mathrm{F}}}:^{-}$$

Note that the sum of these three equations is:

$$\cdot\mathrm{Li} + :\ddot{\underset{\cdot\cdot}{\mathrm{F}}}\cdot \longrightarrow \mathrm{Li}^{+}:\ddot{\underset{\cdot\cdot}{\mathrm{F}}}:^{-}$$

Electrostatic attraction between the positively charged lithium ion and the negatively charged fluoride ion forms the LiF ionic bond. The compound is electrically neutral. Several usual responses help form ionic bonds. Calcium combustion in oxygen produces calcium oxide, a chemical reaction.

$$2Ca\ (s) + O_2(g) \rightarrow 2CaO(s) \tag{4.2}$$

Assuming the dissociation of the diatomic O_2 molecule into individual oxygen atoms, the representation of this reaction can be depicted using Lewis symbols. The examination of the energetics associated with this step will be addressed subsequently.

$$\begin{array}{llcll} \cdot \mathrm{Ca} \cdot \ + & \cdot \ddot{\underset{\cdot\cdot}{\mathrm{O}}} \cdot & \longrightarrow & \mathrm{Ca}^{2+} & :\ddot{\underset{\cdot\cdot}{\mathrm{O}}}:^{2-} \\ [\mathrm{Ar}]4s^2 & 1s^2 2s^2 2p^4 & & [\mathrm{Ar}] & [\mathrm{Ne}] \end{array} \tag{4.3}$$

The calcium atom experiences a transfer of two electrons to the oxygen atom. It is crucial to note that the calcium ion (Ca^{2+}) acquired exhibits the electron configuration characteristic of argon. Furthermore, it is worth noting that the oxide ion (O^{2-}) shares isoelectronic properties with neon. Moreover, the resultant compound, calcium oxide (CaO), preserves electrical neutrality (Reed & Schleyer, 1990).

In several cases, the cation and anion present in a compound display distinct charges. An illustrative instance involves the balanced equation representing lithium combustion in the presence of air, resulting in lithium oxide (Li_2O) formation.

$$4Li(s) + \ O_2(g) \rightarrow 2Li_2O\ (s) \tag{4.4}$$

Using Lewis dot symbols,

$$\begin{array}{llcll} 2\ \cdot \mathrm{Li}\ + & \cdot \ddot{\underset{\cdot\cdot}{\mathrm{O}}} \cdot & \longrightarrow & 2\mathrm{Li}^{+} & :\ddot{\underset{\cdot\cdot}{\mathrm{O}}}:^{2-}\ (\text{or } \mathrm{Li_2O}) \\ 1s^2 2s^1 & 1s^2 2s^2 2p^4 & & [\mathrm{He}] & [\mathrm{Ne}] \end{array} \tag{4.5}$$

In this specific procedure, the oxygen atom experiences reduction by accepting a combined total of two electrons, each derived from one of the two lithium atoms. As a result, the oxygen atom undergoes a transformation, forming the oxide ion. The Li-ion demonstrates isoelectronic properties akin to helium. At higher temperatures, the chemical reaction between magnesium and nitrogen produces a solid compound, magnesium nitride (Mg_3N_2), exhibiting white coloration (Luxford & Bretz, 2014).

$$3Mg\,(s) + N_2(g) \rightarrow Mg_3N_2(s)$$

or

$$\underset{[Ne]3s^2}{3\cdot\mathrm{Mg}\cdot} + \underset{1s^22s^22p^3}{2\cdot\ddot{\mathrm{N}}\cdot} \longrightarrow \underset{[Ne]}{3\mathrm{Mg}^{2+}} \quad \underset{[Ne]}{2:\ddot{\mathrm{N}}:^{3-}} \text{ (or } Mg_3N_2) \qquad (4.7)$$

Each magnesium atom provides two electrons to two nitrogen atoms, transferring six electrons. Mg^{2+} and N^{3-} are isoelectronic with neon. Balanced charges from three Magnesium ions and two Nitrogen ions give the molecule electrical neutrality (Khanna et al., 2002).

Remember

Chemical bonding is not limited to the realm of atoms and molecules on Earth. In space, molecules can form and interact in the interstellar medium, leading to the creation of complex molecules, such as polycyclic aromatic hydrocarbons (PAHs) and amino acids.

5.4. LATTICE ENERGY OF IONIC COMPOUNDS

Even though one can use ionization energy and electron affinity to forecast which elements will form ionic compounds, chemists still study their stability. Gas-phase phenomena determine ionization energy and electron affinity. It is important to note that all ionic compounds are solid at 1 atmospheric pressure and 25°C. In a solid substance, each cation is surrounded by a particular number of anions, and vice versa. Thus, the stability of a solid ionic compound depends on the collective interactions of all constituent ions, not only the interaction between a cation and an anion. Lattice energy quantifies ionic solid stability. The "enthalpy of lattice dissociation" is the energy needed to dissociate one mole of a solid ionic compound into gaseous ions (Kaya et al., 2016).

Direct measurement of lattice energy is not possible. Nevertheless, it is possible to determine the lattice energy of an ionic compound by employing Coulomb's Law, which asserts that the potential energy (E) existing between two ions is directly related to the multiplication of their charges and inversely related to the distance that separates them. The potential energy of a system consisting of a single Lithium-ion and a single Fluoride ion, separated by a distance r, can be expressed as follows (Moody & Thomas, 1965):

$$E \infty Q_{Li} + \frac{Q_F}{r} = k\, Q_{Li} + Q_F/r \tag{4.8}$$

where;and represent the electric charges of the Li and F ions, respectively, whereas 'k' represents the proportionality constant. Due to the positive charge and the negative charge of each of these ions, creating an ionic bond between Li and F is exothermic. Thus, the lattice energy of LiF is positive since energy is needed to reverse the process. Thus, the Li-F bonded pair is more stable than the separate ions.

Indirectly calculating lattice energy implies assuming that an ionic molecule forms sequentially. The Born-Haber cycle links ionic compound lattice energies to atomic and molecular characteristics like ionization energies and electron affinities. The Born-Haber cycle, created by Max Born, a notable German physicist, and Fritz Haber, a distinguished German chemist, describes the stages required to form an ionic solid. (Petrov, 2014).

This experiment calculates lithium fluoride's lattice energy using the above method. This study examines lithium-fluorine chemical reactions. Lithium and Fluorine react together in the following manner:

$$Li(s) + \frac{1}{2}F_2(g) \rightarrow LiF(s) \tag{4.9}$$

This reaction's standard enthalpy change is 2594.1 kJ/mol. The standard enthalpy change for LiF production may be calculated since the reactants and products are in their standard states at 1 atmosphere. The five phases of LiF synthesis from its constituent elements can be examined by applying the fundamental principle that the cumulative enthalpy alterations for each discrete phase are equal to the overall reaction enthalpy (2594.1 kJ/mol). This approach may study energy transformations during ionic compound synthesis, even if the sequence of events may vary. (Yoder & Flora, 2005).

1. Convert solid lithium to lithium vapor (the direct conversion of a solid to a gas is called sublimation):

   $Li(s) \rightarrow Li(g)$

   The energy of sublimation for lithium is 155.2 kJ/mol.

2. Dissociate 1/2 mole of F_2 gas into separate gaseous F atoms:

   $\frac{1}{2}F_2(g) \rightarrow F(g)$

   The energy needed to break the bonds in 1 mole of F_2 molecules is 150.6 kJ. Here, the bonds in half a mole of F_2 are being broken, so the enthalpy change is 150.6/2, or 75.3, kJ.

3. Ionize 1 mole of gaseous Li atoms:

   $Li(g) \rightarrow Li + e$

   This process corresponds to the first ionization of lithium.
4. Add 1 mole of electrons to 1 mole of gaseous F atoms.

   $F(g) + e \rightarrow F(g)$

5. Combine 1 mole of gaseous Li and 1 mole of F to form 1 mole of solid LiF:

   $Li + F \rightarrow LiF$

   The reverse of step 5 is,

   $energy + LiF(s) \rightarrow Li + F$

   This reaction defines the lattice energy of LiF. Thus, the lattice energy must have the same magnitude as H but an opposite sign (Song et al., 2013):

1.	$Li(s) \longrightarrow Li(g)$		$\Delta H_1^\circ = 155.2$ kJ/mol
2.	$\frac{1}{2}F_2(g) \longrightarrow F(g)$		$\Delta H_2^\circ = 75.3$ kJ/mol
3.	$Li(g) \longrightarrow Li^+(g) + e^-$		$\Delta H_3^\circ = 520$ kJ/mol
4.	$F(g) + e^- \longrightarrow F^-(g)$		$\Delta H_4^\circ = -328$ kJ/mol
5.	$Li^+(g) + F^-(g) \longrightarrow LiF(s)$		$\Delta H_5^\circ = ?$
	$Li(s) + \frac{1}{2}F_2(g) \longrightarrow LiF(s)$		$\Delta H^\circ_{overall} = -594.1$ kJ/mol

According to Hess's Law, one can write.

$$\Delta H^o_{overall} = \Delta H_1^0 + \Delta H_2^0 + \Delta H_3^0 + \Delta H_4^0 + \Delta H_5^0 \quad (4.10)$$

or

–59.4 kJ/mol = 155.2 kJ/mol + 75.3 kJ/mol + 520 kJ/mol – 328 kJ/mol +

Hence:

$$\Delta H_5^0 = -1017 kJ/mol$$

and the lattice energy of LiF is 11017 kJ/mol (O'Brien & Akinc, 1990) (Figure 5.2).

Figure 5.2 depicts the Born-Haber cycle for LiF. Steps 1, 2, and 3 require energy allocation. Steps 4 and 5 release energy. The solid LiF's stability depends on its positive lattice energy, which is indicated by its negative H° value. Lattice energy determines ionic compound stability. The endothermic conversion of ions from solid to gaseous states, guided by Coulomb's Law, makes lattice energy always positive.

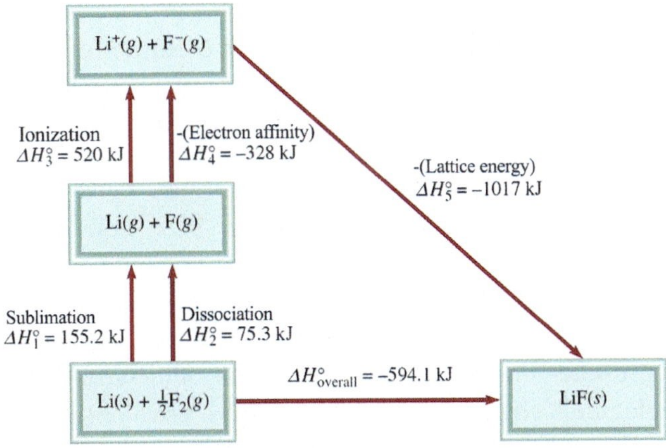

Figure 5.2. *The Born-Haber cycle for the formation of 1 mole of solid LiF.*

Source: Raymond, Creative Commons License.

Table 5.1 contains lattice energies and melting points for many typical ionic compounds. Lattice energy affects the melting point. Lattice energy directly affects solid stability and ionic bond strength. Solid-state melting requires a lot of energy, and hence, materials with greater lattice energies have a higher melting point. $MgCl_2$, MgO, and CaO have greater lattice energies. $MgCl_2$ has an Mg^{2+} cation. In contrast, the other two compounds interact between two species with +2 charges (Mg^{2+} or Ca^{2+}) and a species with –2 charges (O^{2-}). Electrostatic attraction forces are much stronger between species with charges of +2 and –2 than between species with charges of +1 and –1 (Matskevich et al., 2015).

Table 5.1. Lettice Energies and Melting Points of Some Ionic Compounds

Compound	Lattice Energy (kJ/mol)	Melting Point
LiF	1,017	845
LiCl	828	610
NaCl	788	801
NaBr	736	750
$MgCl_2$	2,527	714
MgO	3,890	2,800
CaO	3,414	2,580

5.5. THE COVALENT BOND

Chemists did not fully comprehend molecular creation until the early 20th century, even though molecules were first discovered in the 17th century. Gilbert Lewis proposed that chemical bonding involves atoms sharing electrons.

$$H\cdot + \cdot H \rightarrow H{:}H$$

Electron pairing illustrates a covalent connection involving two atoms sharing two electrons. Covalent substances have only covalent bonding. A single line represents the covalently coupled pair of electrons for clarity. The hydrogen molecule's covalent link is H→H. In a covalent bond, the shared pair of electrons attract the nuclei of both atoms. Hydrogen bonds are the attractive connection between two atoms in an H_2 molecule. This bond helps form covalent connections in many compounds (Schomaker & Stevenson, 1941). Valence electrons dominate covalent bonding between multi-electron atoms. This investigation will examine fluorine's molecular structure (F_2). Fluorine's electron configuration is $1s^2\,2s^2\,2p^5$. The 1s orbital electrons have a low energy level and tend to stay close to the atomic nucleus. Because of this, they avoid bonding. Thus, each fluorine atom has seven valence electrons in 2s and 2p orbitals. Thus, fluorine gas (F_2) has the following molecular structure:

$$:\ddot{\underset{..}{F}}\cdot + \cdot\ddot{\underset{..}{F}}: \longrightarrow :\ddot{\underset{..}{F}}:\ddot{\underset{..}{F}}: \quad \text{or} \quad :\ddot{\underset{..}{F}}-\ddot{\underset{..}{F}}:$$

It should be noted that the formation of F_2 involves the participation of only two valence electrons. The valence electrons not engaged in forming covalent bonds are referred to as lone pairs, which consist of unbonded pairs of electrons. Therefore, it can be observed that each atom of F in the compound F_2 possesses a total of three unshared pairs of electrons (Huang et al., 2020):

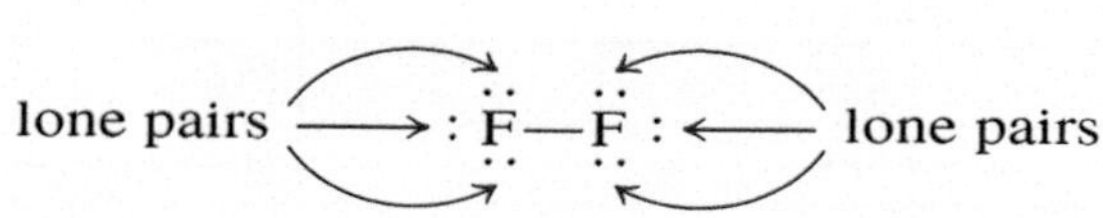

The terminology employed to depict covalent compounds, exemplified by H_2 and F_2, is called Lewis structures. A Lewis structure is a graphical depiction of covalent bonding, illustrating the sharing of electron pairs between two atoms. Either lines or pairs of dots represent shared electron pairs, while lone pairs are depicted as pairs of dots located on individual atoms. In a Lewis structure, solely the valence electrons are depicted. Following is the explanation of the Lewis structure of the water molecule. The Lewis dot representation of oxygen illustrates the presence of two unpaired dots or two unpaired electrons. This observation aligns with the expectation that oxygen has the potential to engage in the formation of two covalent bonds. Due to the presence of a single electron, hydrogen can form solely one covalent bond. Therefore, the Lewis structure representing the molecular arrangement of water is as follows (Feinberg & Ruedenberg, 1971):

$$H:\ddot{\underset{\cdot\cdot}{O}}:H \qquad \text{or} \qquad H-\ddot{\underset{\cdot\cdot}{O}}-H$$

In this particular scenario, the oxygen atom possesses two unshared electron pairs. The absence of lone pairs in the hydrogen atom can be attributed to using its sole electron to establish a covalent bond. The attainment of a noble gas configuration in the F_2 and H_2O molecules occurs through the process of electron sharing:

These molecules demonstrate Lewis's octet rule. Until they have eight valence electrons, atoms, except hydrogen, tend to bind. A covalent bond forms when each atom lacks enough electrons to fill an octet. Atoms establish covalent bonds by sharing electrons, completing their valence shells, and satisfying the octet rule. Hydrogen requires the two-electronic configuration of helium (Tichý & Ticha, 1995).

The octet rule is most common in second-period elements. These elements only contain 2s and 2p subshells, which have eight electrons. When these atoms create a covalent combination, they can share electrons to achieve the electron configuration of a noble gas [Ne]. The next sections will investigate prominent exceptions from the octet rule to understand chemical bonding better. Atoms can form many covalent bonds. Two atoms share electrons in a covalent link. Multiple compounds are connected when two atoms share two or more electron pairs. A "double bond" is a covalent link between two atoms sharing two electron pairs. Double bonds make up CO_2 and C_2H_4 molecules (Beier & Hoheisel, 1999) (Figure 5.3).

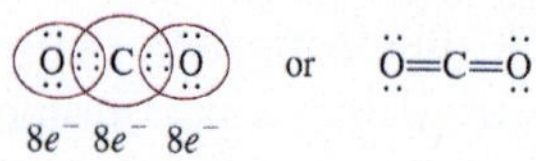

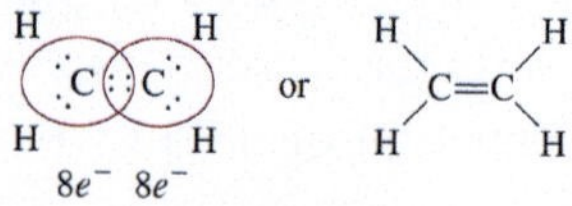

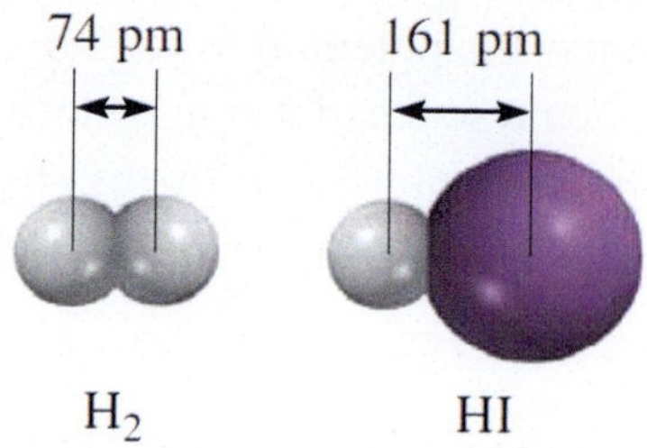

Figure 5.3. *The bond length in pm in H_2 and HI.*

Source: McCarty, Creative Commons License.

A triple bond arises when two atoms share three pairs of electrons, as in the nitrogen molecule (N_2):

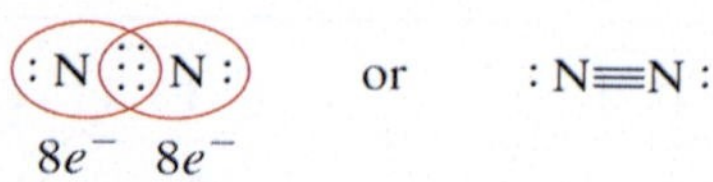

The acetylene molecule (C_2H_2) also contains a triple bond, in this case between two carbon atoms:

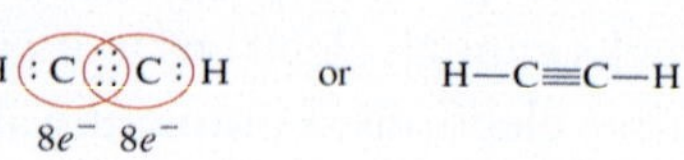

In ethylene and acetylene, all valence electrons bind; hence the carbon atoms have no non-bonding electron pairs. Unlike carbon monoxide, most stable carbon-based compounds lack lone pairs on their carbon atoms. Multiple covalent connections between atoms diminish bond length. Bond length quantifies the interatomic spacing between the nuclei of two covalently bonded atoms in a molecule (de Oteyza et al., 2013).

Did You Know?

The strongest chemical bond is the carbon-carbon triple bond found in acetylene (C2H2). This bond is so strong that it releases a significant amount of energy when broken, making acetylene an important fuel in welding torches.

As said, atoms share electron pairs to form a covalent connection. In a chemical system like H_2, when the atoms are indistinguishable, electrons should have a uniform distribution and identical closeness to each atom. Due to the differences between hydrogen (H) and fluorine (F) atoms, the HF molecule, which forms a covalent connection, distributes bonding electrons unequally (Hinze & Jaffe, 1962).

HF has a polar covalent bond. This terminology comes from the uneven electron density of the bonding atoms. Based on actual data, the electrons in the HF molecule concentrate around the fluorine (F) atom. Scientific literature often refers to the difference in electron allocation between hydrogen and fluorine as a partial electron transfer or electron density redistribution. "Asymmetric sharing" of the bonding electron pair increases electron density near the fluorine atom but decreases near hydrogen (Parr & Pearson, 1983).

Polar bonds like hydrogen-fluorine (HF) are midway between nonpolar covalent connections, where electrons are exchanged evenly, and ionic bonds, where electrons are virtually entirely transported. Atoms attract electrons in chemical bonds to their nucleus due to their electronegativity. This feature helps distinguish polar covalent bonds from nonpolar ones. High-electronegativity elements attract electrons better than low-electronegativity ones. Electronegativity is likely related to electron affinity and ionization energy. Thus, an element like fluorine, which has a high electron affinity and ionization energy, has a high electronegativity. Sodium has lower electron affinity, ionization energy, and electronegativity (Pearson, 1988) (Figure 5.4).

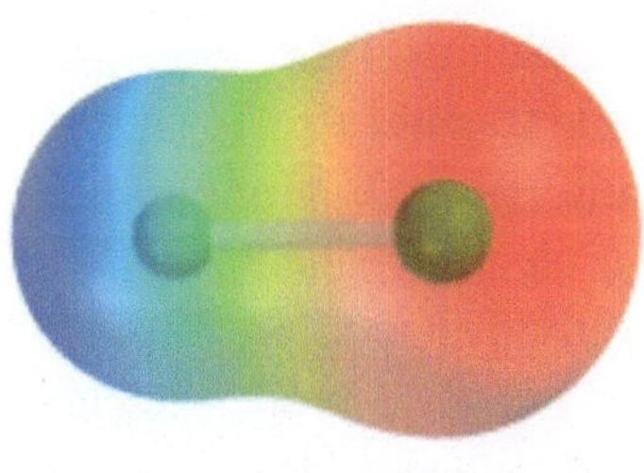

Figure 5.4. *Electrostatic potential map of the HF molecule. The distribution varies according to the colors of the rainbow. The most electron-rich region is red; the most electron-poor region is blue.*

Source: Harald, Creative Commons License.

Electronegativity necessitates comparison to other elements. An eminent American scientist, Linus Pauling, devised the method for detecting elemental electronegativities. After examining the figure, there are clear patterns and relationships between element electronegativity levels. From left to right in a periodic table period, electronegativity rises. This reduces elements' metallic characteristics. As atomic number increases, electronegativity decreases, and metallic characteristics increase within each group. Transition metals deviate from trends. The top right quadrant of the periodic table contains the halogens, oxygen, nitrogen, and sulfur, which have the highest electronegativity. Alkali and alkaline earth metals, with the lowest electronegativity, tend to cluster in the lower left of the periodic table (Batsanov, 1982) (Figure 5.5).

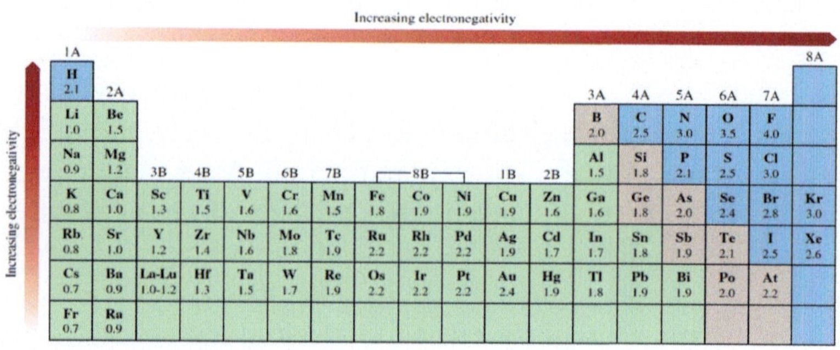

Figure 5.5. *The electronegativity of common elements.*

Source: Walsh, Creative Commons License.

NaCl and CaO are examples of ionic bonding between elements with varying electronegativities. The atom of the element with lower electronegativity prefers to transfer its electron(s) to the element with higher electronegativity. An ionic connection forms when a metallic and nonmetallic atom combine. Because electron density displacement is small, elements with similar electronegativities tend to form polar covalent connections. Covalent bonds develop when nonmetallic atoms share electrons. Only atoms of the same element with similar electronegativity form pure covalent bonds. Ionization energies and electron affinities match the observed trends (Miedema et al., 1973) (Figure 5.6).

The following principles are a good estimate for distinguishing polar covalent bonds from ionic bonds. An ionic connection occurs when the electronegativity difference between two atoms is 2.0 or larger.

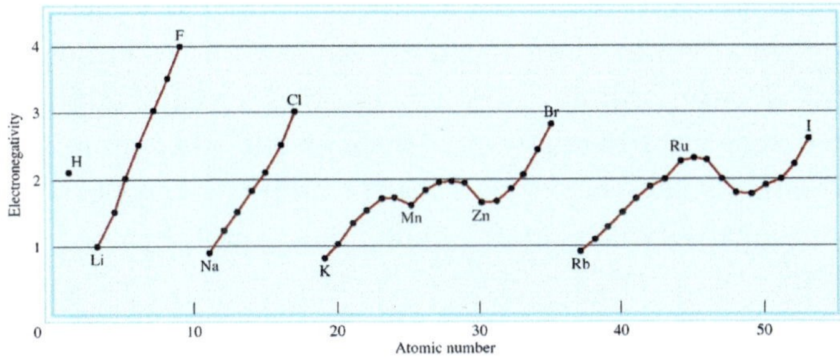

Figure 5.6. *Variation of electronegativity with atomic number. The halogens have the highest electronegativities, and the alkali metals have the lowest.*

Source: David, Creative Commons License.

Exceptions occur, but most ionic substances follow this, Law. 0.3 to 2.0 electronegativity differences between atoms generate polar covalent bonds. The bond is non-polar if the difference in electronegativity between two atoms is below 0.3. Chemists may define chemical bonds using the percent ionic character. Ionic bonds are 100% ionic. No such link has been empirically observed. The H_2 molecule, for example, has no ionic properties (Noorizadeh & Shakerzadeh, 2008).

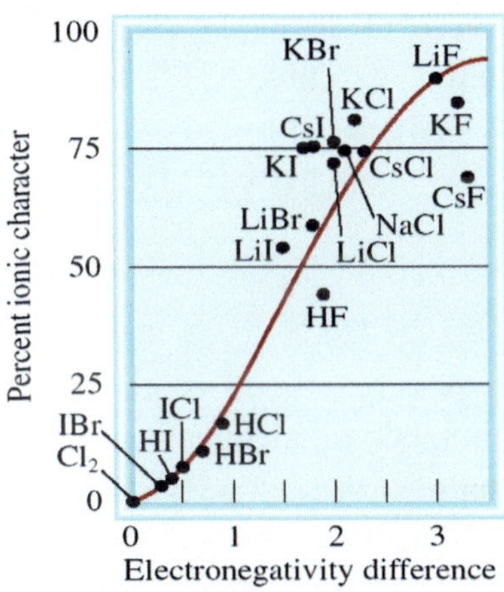

Figure 5.7. *Relation between percent ionic character and electronegativity difference.*

Source: Peter, Creative Commons License.

Electronegativity and electron affinity are related topics in chemistry. Both events involve an atom's natural inclination to attract electrons to its nucleus. Electron affinity is the tendency of a single atom to attract an electron, whereas electronegativity is the capacity of an atom in a chemical bond to attract the shared electrons. Electron affinity is physically observable, while electronegativity is theoretical and unmeasurable (Reddy et al., 1998) (Figure 5.7).

Did You Know?

Water's hydrogen bonding plays a crucial role in many biological processes. It gives water its high heat capacity, enabling it to regulate temperatures in living organisms and maintain stable environments for life.

5.6. ELECTRONEGATIVITY AND OXIDATION NUMBER

The oxidation number is the theoretical charge an atom would have if electrons were transported to a molecule's more electronegative atom. The nitrogen atom in the NH_3 molecule creates three covalent connections with three hydrogen atoms. Nitrogen's stronger electronegativity transfers electron density from hydrogen to it. If the transfer is complete, each hydrogen atom (H) will give an electron to nitrogen (N), giving nitrogen a net charge of +3 and each hydrogen atom a –1. Thus, NH_3's nitrogen oxidation state is +3, while hydrogen's is –1. Oxygen has a –2 oxidation number in most chemical compounds. However, hydrogen peroxide (H_2O_2) has a Lewis structure like this (Duffy & Ingram, 1975):

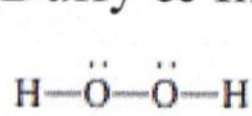

Since the electron pair in a bond is uniformly divided amongst the atoms, the oxidation number of identical atoms does not change. Since hydrogen's oxidation number is 1, oxygen's is –2. Do you understand why fluorine always has a –1 oxidation number? This element is noted for having the greatest electronegativity. Additionally, its molecules often form single chemical bonds. Thus, full electron transfer yields a charge of –1 (Arakawa et al., 2018).

The octet rule and Lewis structures do not fully describe covalent bonding, but they help explain molecular characteristics and reactions by showing bonding patterns in diverse compounds. Because of this, building

Lewis structures for chemical compounds is advised. The basic steps are (Finkenstaedt-Quinn et al., 2019):

1. The task of identifying simple compounds is relatively straightforward. For more intricate compounds, obtaining the relevant information or employing a reasoned inference is necessary. Typically, the central position is occupied by the atom with the lowest electronegativity. In the Lewis structure, it is customary for hydrogen and fluorine to occupy the terminal positions typically (Lever, 1972).
2. To determine the total charge of polyatomic anions, one must add the number of negative charges to the existing total. In the case of the CO_2 ion, adding two electrons is warranted due to the 2-charge, which signifies an excess of two electrons beyond the number contributed by the constituent atoms. In the case of polyatomic cations, the total charge is obtained by subtracting the number of positive charges. Therefore, in the case of NH_4^+, the deduction of one electron is made due to the presence of a charge of +1, which signifies the removal of one electron from the collective assembly of atoms.
3. Create a sole covalent bond connecting the central atom with each adjacent atom. Ensure that the atoms bonded to the central atom have complete octets. It is important to note that the valence shell of a hydrogen atom achieves stability with a total of two electrons. Lone pairs of electrons must be depicted for central or surrounding atoms when they are not engaged in bonding. The quantity of electrons to be utilized is ascertained in step 2 (Ahmad & Zakaria, 2000).
4. Upon the conclusion of steps 1–3, if the central atom possesses an electron count below eight, it is advisable to consider incorporating double or triple bonds between the neighboring and central atoms. This can be achieved by utilizing lone pairs originating from the surrounding atoms, thereby fulfilling the octet requirement of the central atom (Packer & Woodgate, 1991).

Did You Know?

A solitary DNA molecule can sustain a 1,000-fold stress on its covalent link. Given the DNA molecule's small size and importance in genetic data retention and diffusion, the achievement is remarkable.

5.7. FORMAL CHARGE AND LEWIS STRUCTURE

By conducting a comparison between the electron count of an atom in an isolated state and its electron count in a Lewis structure, it becomes feasible to determine the electron distribution of a molecule and construct the Lewis structure that is most probable. The number of electrons in an atom is equivalent to the number of valence electrons it possesses. Typically, inner electrons are deemed excessive. Within a molecular system, the electrons associated with an atom encompass both bonding and nonbonding electrons. The equal sharing of electrons is a requirement for bonding pair atoms due to electron sharing. The concept of formal charge in atomic systems pertains to the difference in electric charge between the valence electrons of an isolated atom and the electrons present in its corresponding Lewis structure (Purser, 2001). The following steps are utilized to ascertain the number of electrons bonded to an atom within a Lewis structure.

All the atom's nonbonding electrons are assigned to the atom. The bond(s) between the atom and other atoms(s) are broken and are assigned half of the bonding electrons to the atom.

This concept is illustrated using the ozone molecule (O_3).

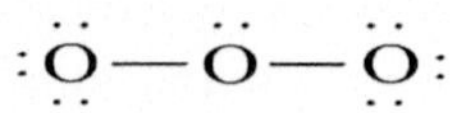

It is evident that despite utilizing all accessible electrons, the central atom fails to adhere to the octet rule. This issue is addressed by transforming a non-bonding electron pair located on one of the terminal atoms into a second covalent bond connecting said terminal atom with the central atom in the following manner (Curnow & Pearson, 2023):

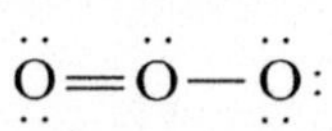

The formal charge on each atom in O_3 can now be calculated according to the following scheme:

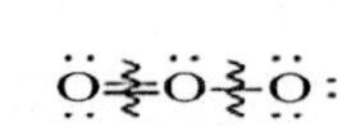

	O	O	O
Valence e^-	6	6	6
e^- assigned to atom	6	5	7
Difference (formal charge)	0	+1	−1

The wavy red lines represent the disruption of chemical bonds. It should be noted that the rupture of a singular bond leads to the transfer of one electron, while the rupture of a double bond results in the transfer of two electrons to each of the atoms involved in the bonding, and so forth. Hence, the formal charges assigned to the atoms in the O_3 molecule are (See, 2009):

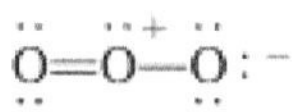

In the case of individual positive and negative charges, it is customary to exclude the numerical value of 1. When composing formal charges, the following guidelines can be beneficial (Cooper et al., 2012).

In the context of molecular entities, the summation of formal charges must equate to zero, as molecules inherently possess electrical neutrality. (This rule is applicable, exemplified by the O_3 molecule).

- In the case of cations, it is necessary for the sum of the formal charges to be equivalent to the positive charge.
- In the case of anions, the total of the formal charges must be equivalent to the negative charge (Karonen et al., 2021).
- It is important to note that formal charges do not accurately depict the true distribution of charges within a molecule. In the O_3 molecule, there is a lack of evidence supporting the notion that the central atom carries a net charge of 11 or that one of the end atoms carries a charge of 21. The assignment of charges to the atoms in the Lewis structure facilitates the organization and monitoring of the valence electrons within the molecule (McGoran, 1991).
- In certain cases, it is possible to encounter multiple valid Lewis structures for a given chemical species. In instances of this nature, it is frequently possible to determine the most credible Lewis's structure by employing formal charges and adhering to the subsequent principles (McArdle, 2019).
- In the context of molecular structures, having a Lewis structure devoid of formal charges is generally more favorable than one that exhibits formal charges (Reiher & Sundermann, 2002).
- Lewis structures that exhibit high formal charges, such as 12, 13, 22, 23, and so forth, are deemed less probable compared to those with lower formal charges (Shelley et al., 2007).
- The Lewis structure that exhibits the most credibility among those with comparable distributions of formal charges is the structure wherein negative formal charges are allocated to the atoms with higher electronegativity (Habiddin et al., 2023).

5.8. THE CONCEPT OF RESONANCE

The Lewis structure representation of ozone (O_3) adhered to the octet rule for the central atom by establishing a double bond between the central atom and one of the two terminal oxygen atoms. Indeed, it is possible to position the double bond at either terminus of the molecule, as evidenced by the presentation of these two isomeric Lewis structures that possess identical properties (Bruton & Bartley, 1985):

Nevertheless, both Lewis structures fail to explain the observed bond lengths in O_3. It is anticipated that the O-O bond in O_3 would exhibit greater lengths than the O-P bond due to the established tendency of double bonds to possess shorter lengths than single bonds. However, empirical data indicates that the lengths of both oxygen-to-oxygen bonds are equal, measuring 128 picometers. The discrepancy is resolved by employing both Lewis structures as a means to depict the ozone molecule (Susen, 2020):

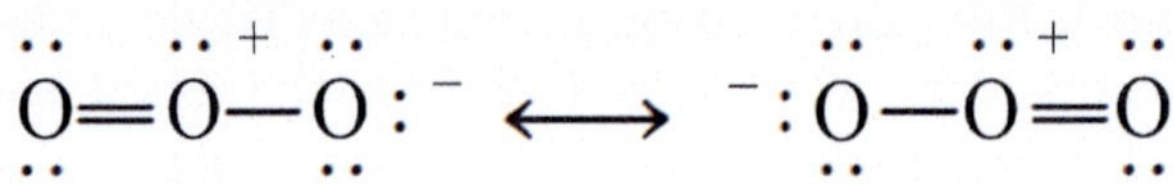

Each of these structures is referred to as a resonance structure. A resonance structure refers to one of multiple Lewis structures representing a single molecule, as a single Lewis structure cannot accurately depict its structure. The presence of a double-headed arrow signifies that the depicted structures are resonance structures. "resonance" refers to utilizing multiple Lewis structures to depict a specific molecule (Antonov, 2010). Similar to the accounts of medieval European travelers who encountered rhinoceroses in Africa and described them as a hybrid creature combining the features of a griffin and a unicorn, one can employ a comparable approach when characterizing ozone, an actual molecule, by utilizing two conceptual structures that are familiar yet non-existent. A prevalent misinterpretation regarding resonance pertains to the mistaken belief that a molecule, such as ozone, undergoes rapid oscillation between its different resonance structures. It is important to note that neither of the resonance structures presented can fully depict the true structure of the molecule, as it possesses its own distinct and stable arrangement. The concept of "resonance" is a construct developed by humans to overcome the constraints inherent in simplistic bonding models. To further elaborate on the animal analogy, it is important to note that a rhinoceros possesses unique characteristics and should not be perceived as a mere fluctuation between the mythical creatures of a griffin and a unicorn (Goodman et al., 1996).

The carbonate ion serves as an additional illustration of resonance:

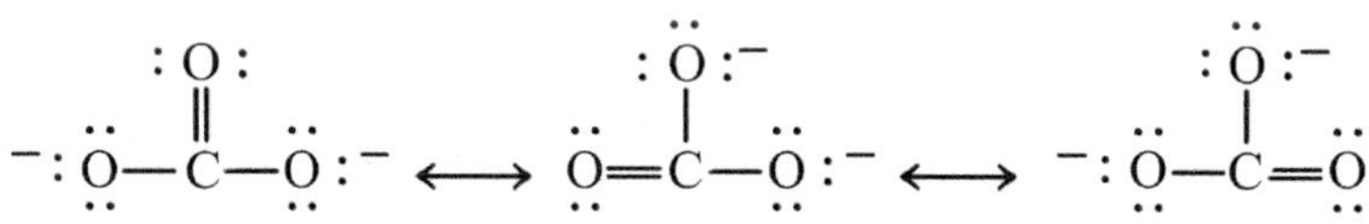

Based on empirical findings, it has been observed that the carbon-to-oxygen bonds within the carbonate ion (CO_3^{2-}) exhibit equivalence. Hence, a comprehensive understanding of the carbonate ion's characteristics necessitates considering its collective resonance structures. The application of the concept of resonance is equally relevant to organic systems. An illustrative instance can be observed in the benzene molecule (C_6H_6) (Xue & Stains, 2020):

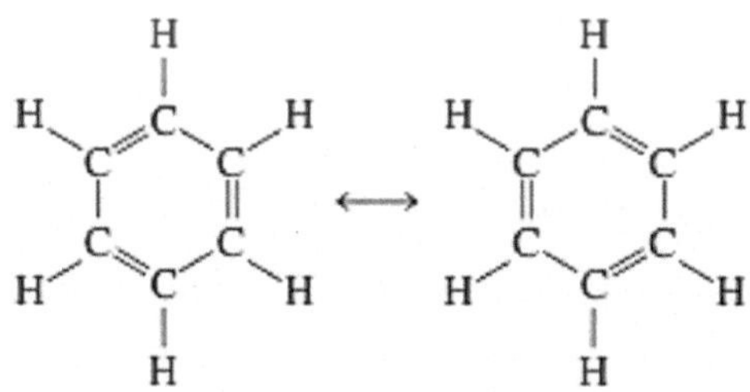

If one of these resonance structures were to represent benzene's molecular structure accurately, it would result in two distinct bond lengths between adjacent carbon atoms. These bond lengths correspond to the characteristics of single and double bonds, respectively. Indeed, it is worth noting that the interatomic distance between adjacent carbon (C) atoms in benzene is measured to be 140 picometers (pm). This distance is comparatively shorter than that of a carbon-oxygen-carbon (COC) bond, which spans 154 pm, while it is longer than that of a carbon-phosphorus-carbon (CPC) bond, which extends over a distance of 133 pm. A more streamlined approach to illustrating the molecular structure of benzene and other compounds featuring the "benzene ring" involves solely depicting the skeletal framework, omitting the explicit representation of carbon and hydrogen atoms. In accordance with this convention, the representation of resonance structures is as follows (Weishaupt et al., 2006):

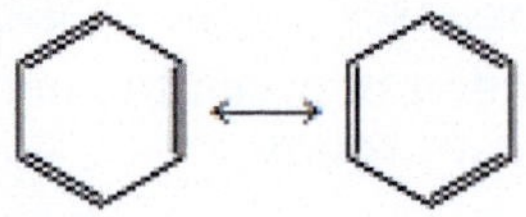

It is important to acknowledge that the carbon atoms located at the corners of the hexagon and the hydrogen atoms have been excluded from the representation, albeit their presence is implied. The molecular diagram only depicts the covalent bonds formed between the carbon atoms. It is imperative to bear in mind the following crucial principle when generating resonance structures: Different resonance structures allow for the rearrangement of electrons, specifically bonds, while the positions of atoms remain unchanged. However, the requirement is that identical atoms form bonds in all resonance structures of a particular species (Keller, 2005).

Remember

Ionic bonding is responsible for the bright colors of many gemstones. For example, the vibrant blue color of sapphires comes from the presence of titanium and iron impurities, which create charge imbalances in the crystal lattice.

5.9. METALLIC BONDING

Metallic bonding is an intriguing and distinctive form of chemical bonding that governs the behavior of metals, thereby influencing their characteristic properties, including malleability, ductility, and electrical conductivity. In contrast to alternative forms of bonding, metallic bonding entails the distribution of electrons across a considerable number of atoms, forming a cohesive and exceptionally conductive lattice structure. The phenomenon of bonding plays a crucial role in comprehending the behavior of metals and their extensive applications across diverse industries (Acar & Tarhan, 2008).

The fundamental basis of metallic bonding is rooted in the organization of electrons within metal atoms. Metals exhibit a limited number of valence electrons within their outermost energy level, resulting in a scenario where these electrons experience a relatively weak attraction from the atomic nucleus. Consequently, the valence electrons undergo delocalization, thereby acquiring the ability to migrate across the entirety of the metal lattice freely. The distinguishing characteristic of metallic bonding from ionic or covalent bonding is the localization of electrons between specific atoms in the latter two types (Gade, 2000).

Within a metallic lattice, the atomic cores, which bear a positive charge and are referred to as cations, are encompassed by a collective of electrons that can move freely throughout the lattice, forming what is commonly known as a "sea" of delocalized electrons. The electron cloud exhibits a homogeneous distribution of negative charge, which maintains the structural integrity and stability of the metal atoms. The malleability and ductility of the metal lattice can be attributed to the lack of localized electrons, which enables the cations to undergo positional shifts within the lattice without causing any disturbance to the electron cloud (Gou et al., 2008).

The enhanced electrical and thermal conductivity observed in metals can be attributed to the high degree of electron mobility facilitated by metallic bonding. Applying an electric field to a metal induces the movement of delocalized electrons in response to the electric potential, thereby generating an electric current. Likewise, when exposed to heat, the mobile electrons effectively transmit thermal energy across the metal lattice, resulting in a notable level of thermal conductivity (Collman & Boulatov, 2002).

The robustness of metallic bonding results from the potent electrostatic attraction between the positively charged metal cations and the mobile, or delocalized, electrons. The phenomenon responsible for metals' high melting and boiling points results from the significant energy necessary to disrupt the cohesive forces between the metal ions and the electron cloud. Furthermore, metallic bonding confers metals with a high density due to the tightly packed lattice structure, which maximizes the number of atoms within a given volume (Terada et al., 2002).

One fascinating outcome of metallic bonding is the "electron sea" conductivity phenomenon. The movement of electrons within the lattice facilitates their interaction with photons, resulting in light energy absorption and subsequent reemission. The phenomenon mentioned above imparts metals with their distinctive luster and reflective attributes, resulting in their shiny and metallic appearance (Seshadri et al., 1998).

Metals play a crucial role in various applications within contemporary society, encompassing both industrial and consumer domains. The indispensable role of metals in constructing electrical wires, circuits, and electronic devices stems from their inherent property of electrical conductivity. The inherent malleability of the material facilitates its ability to be molded into diverse configurations, thereby facilitating the creation of items such as coins, foils, and sculptures. Furthermore, the capacity of metallic bonding to endure stress and exhibit elastic behavior renders it indispensable in the fabrication of structural elements in architecture and engineering (Gervasio et al., 2004).

Comprehending metallic behavior and properties via metallic bonding has also facilitated the advancement of sophisticated materials. Alloying is a metallurgical procedure involving combining multiple metals, resulting in modified metal properties that can be tailored to meet specific requirements. These modifications may include strength enhancements, corrosion resistance, or magnetic characteristics. Illustrative instances encompass stainless steel, a composite material that integrates chromium to augment its resistance against corrosion, and bronze, a compound that amalgamates copper with tin to enhance its overall durability (Filippetti & Fiorentini, 2005).

Although metallic bonding is utilized in various fields, it is important to acknowledge its inherent limitations. The presence of fixed and immobile cation cores imposes constraints on the reactivity of metals. While this can offer benefits such as enhancing the stability and integrity of the metal, it can also pose limitations in situations where reactivity is sought after (Gould et al., 2022).

In summary, metallic bonding is an extraordinary phenomenon that distinguishes metals from other substances. The cohesive and conductive lattice in metals arises from the sharing of delocalized electrons, which is responsible for their distinctive characteristics, including electrical conductivity, malleability, and luster. The phenomenon of metallic bonding is of utmost significance in shaping the physical landscape of the world, encompassing various domains such as infrastructure, technology, and the development of sophisticated materials. As the investigation and comprehension of metallic bonding progress, novel materials and technological advancements emerge, presenting transformative prospects poised to influence future developments significantly (Schultz et al., 2005).

5.10. PHASE CHANGES AND ENERGY ABSORPTION

The phenomenon of phase changes, including evaporation, condensation, melting, and freezing, is significantly influenced by intermolecular forces. A substance must either surpass or establish intermolecular forces to undergo a phase transition. During evaporation, thermal energy disrupts the intermolecular forces that maintain cohesion among molecules within a liquid, thereby enabling their transition into a gaseous state (Khudhair & Farid, 2021).

5.11. SOLUBILITY AND COLLIGATIVE PROPERTIES

The comprehension of solubility, which refers to the capacity of one substance to dissolve in another, is contingent upon a thorough understanding of intermolecular forces. Substances exhibiting comparable intermolecular forces have a higher propensity to dissolve mutually. Polar solutes tend to dissolve in polar solvents, whereas nonpolar solutes demonstrate enhanced solubility in nonpolar solvents.

The influence of intermolecular forces extends to colligative properties, including boiling point elevation and freezing point depression. Introducing solute particles into a solvent causes a disturbance in the solvent's organized structure, leading to alterations in the boiling and freezing points (Proniuk & Blanchard, 2001).

5.12. MOLECULAR ORBITAL THEORY

Quantum chemistry's Molecular Orbital (MO) Theory explains electron behavior in molecular systems. The theory of molecular orbitals (MO) expands classical theories of molecule characteristics and interactions by applying quantum mechanics. This theory explains molecules' electronic structure, bonding, and spectroscopic characteristics, revealing the molecular domain's complicated quantum qualities.

Electron wave-like features underlie molecular orbital (MO) theory. As the Bohr model suggests, quantum mechanics shows that electrons do not orbit the nucleus in stationary orbits. Atomic orbitals reflect the chance of finding an electron in an area around the nucleus.

In MO theory, atomic orbitals from different atoms converge to form molecular orbitals. When two atoms create a molecule, their atomic orbitals merge and undergo constructive and destructive interference. Constructive interference creates bonding molecular orbitals (MOs), which help electrons localize between nuclei. Localization increases atom chemical bond formation. Destructive interference creates antibonding molecular orbitals (MOs) with lower electron probability density in the interatomic region. This destabilizes the molecule (Hehre et al., 1970).

Linear combination of atomic orbitals (LCAO) may mathematically characterize molecular orbital generation. LCAO combines atomic orbitals by addition or subtraction to create bonding and antibonding molecular orbitals (MOs). The linear combination coefficients of atomic orbitals determine the shape and energy of the molecular orbitals.

MO theory predicts molecule stability and energy levels. Atomic orbitals and their kinds and amounts determine molecular orbital energy levels. Bonding molecular orbitals (MOs) have lower energy levels than their atomic orbitals. Antibonding MOs have higher energies. Bond dissociation energy is often used to explain the energy difference between bonding and antibonding chemical orbitals. This metric quantifies chemical bond disruption energy.

Molecular Orbital (MO) theory helps explain chemical bond stability and resilience. Electrons first occupy bonding molecular orbitals (MOs) with lower energy levels in a thermodynamically stable molecular system before moving to higher-energy antibonding MOs. Electrons in molecular orbitals produce cohesive forces that bond atoms. Electrons in antibonding molecular orbitals (MOs) reduce the bond intensity, which may cause dissociation (Pople et al., 1965).

Molecular orbital theory explains molecule forms and characteristics. Quantity and energy levels of bonding and antibonding molecular orbitals can determine a molecule's bond order. The bond order may determine the relative bond strength. Higher bond orders are stronger. Bond orders 1 and 2 imply single and double bonds, respectively.

MO theory accurately predicts molecular shapes and geometry. Sigma and pi bonds, which result from molecular orbital (MO) overlap, explain the molecular structures of ethene (C_2H_4) and benzene (C_6H_6). The theory explains how non-bonding electron pairs affect molecular geometries and bond angles in substances like water (H_2O) and ammonia (NH_3) (Suzuki, 2012).

MO theory's ability to predict molecules' spectroscopic characteristics is important. Molecular orbital transitions control light absorption and emission. Each molecule's electronic spectra are determined by the wavelengths of light absorbed or emitted by these orbitals' energy levels. By examining these spectra, scientists may detect functional groups and understand molecular electronic configurations.

MO theory is essential in computational chemistry and materials research due to its adaptability and predictability. Researchers use molecular orbital (MO) theory to study and create compounds with customized application features, including pharmaceutical development, catalyst design, and material production.

Although MO theory has achieved notable successes, it has limitations. The methodology depends on estimations and reductions; in the case of intricate systems, the computations may require significant computational resources. Furthermore, it should be noted that molecular orbital (MO) theory may not comprehensively account for the influence of electron correlation, wherein the motion of a single electron can impact the behavior of other electrons. Due to this rationale, additional theoretical approaches, such as density functional theory (DFT), are frequently utilized with molecular orbital (MO) theory to yield more precise outcomes for larger molecules and systems (Longuet - Higgins, 1950).

In summary, Molecular Orbital Theory offers a comprehensive detail of the quantum characteristics of electrons within molecules, thereby unveiling the intricate aspects of chemical bonding and molecular properties. Utilizing molecular orbital (MO) theory allows for predicting molecular shapes, bond strengths, and spectroscopic properties by considering the wave-like characteristics of electrons and the interactions among atomic orbitals. The above-mentioned theory has significantly transformed our understanding of molecular dynamics, serving as a guiding principle for scientific investigations across various disciplines, including chemistry, physics, biology, and materials science. With the development of computational capabilities, the utilization of MO theory is experiencing a broadening scope, enabling the exploration of novel avenues in molecular design and pushing the limits of the comprehension of the molecular realm. Intermolecular forces are universally present and play a fundamental role in the comprehension of molecular phenomena. From the least potent London dispersion forces to the more robust and specific hydrogen bonding, these forces play a significant role in shaping the properties and interactions of molecules in various manners. Intermolecular forces are important in diverse academic disciplines, encompassing chemistry, physics, biology, and materials science. Through examining and utilizing these forces, researchers persist in uncovering novel perspectives and practical implementations that enhance the comprehension of substance and propel technological progress (Fukui et al., 1954).

Remember

Chemical bonding plays a vital role in the functioning of electronic devices. The precise arrangement of atoms and their bonds in semiconductors, such as silicon, enables the manipulation of electrons, making modern electronics possible.

5.13. SUMMARY

Atoms create molecules and compounds by chemical bonding. Many theories and models have been created to comprehend chemical bonding and molecular characteristics. This chapter examines bonding theories and their chemical applications.

Lewis dot symbols clearly illustrate atoms' valence electrons in chemical processes. Lewis dots are placed around the chemical symbol to identify valence electrons and predict chemical bonding patterns.

Electron transport between atoms creates ions with opposing charges. Due to strong lattice forces, electrostatic interaction between positive cations and negative anions creates ionic compounds with high melting and boiling temperatures. The energy needed to separate ions is used to quantify lattice energy in ionic compounds, which affects their stability.

Atoms exchange electrons to create covalent bonds. Chemical bonding creates molecular complexes with unique geometries and properties. Electronegativity indicates an atom's capacity to attract shared electrons in a covalent connection, which affects bonding types. Oxidation numbers help explain electrons and charge distribution in molecules and polyatomic ions.

Formal charge and Lewis structures help identify molecular systems' most stable atom and electron arrangements. Formal charges on atoms help find the best Lewis structures for molecules. However, resonance delocalizes electrons across several atoms, allowing certain compounds to have numerous valid Lewis structures. Resonance strengthens molecules, especially aromatic compounds.

Metallic bonding forms a coherent lattice structure with metal cations surrounded by a "sea" of electrons. They have high electrical and thermal conductivity. Metals' malleability and ductility are caused by this bonding.

Boiling, melting, surface tension, and capillary action are affected by intermolecular forces. "Forces" are molecules' attractive interactions. London dispersion forces, dipole-dipole interactions, and hydrogen bonding are included. Hydrogen bonding shapes water and biomolecules.

Quantum mechanics-based Molecular Orbital (MO) Theory explains electronic behavior in molecular systems. Combining atomic orbitals, molecular orbital (MO) theory explains molecule geometries, bond energies, and spectroscopic properties. This technique can accurately predict molecular characteristics and aid pharmaceutical and material science research. Chemical bonding and molecular structure must be analyzed to comprehend atomic and molecular characteristics. Lewis dot symbols, ionic bonding, covalent bonding, and molecular orbital theory help understand chemical bonds and substances. Intermolecular forces are crucial to understanding liquid and solid characteristics. Resonance and formal charge both affect molecular stability and reactivity. These notions underpin contemporary chemistry and advance many scientific and technological fields.

REFERENCES

1. Acar, B., & Tarhan, L., (2008). Effects of cooperative learning on students' understanding of metallic bonding. *Research in Science Education, 38*, 401–420.
2. Ahmad, W. Y., & Zakaria, M. B., (2000). Drawing Lewis structures from Lewis symbols: A direct electron-pairing approach. *Journal of Chemical Education, 77*(3), 329.
3. Antonov, A. A., (2010). New interpretation of resonance. *International Journal of Pure and Applied Sciences and Technology, 1*(2), 1–12.
4. Arakawa, M., Ando, K., Fujimoto, S., Mishra, S., Patwari, G. N., & Terasaki, A., (2018). The role of electronegativity on the extent of nitridation of group 5 metals as revealed by reactions of tantalum cluster cations with ammonia molecules. *Physical Chemistry Chemical Physics, 20*(20), 13974–13982.
5. Batsanov, S. S., (1982). Dielectric methods of studying the chemical bond and the concept of electronegativity. *Russian Chemical Reviews, 51*(7), 684.
6. Beier, M., & Hoheisel, J. D., (1999). Versatile derivatization of solid support media for covalent bonding on DNA-microchips. *Nucleic Acids Research, 27*(9), 1970–1977.
7. Bruton, L., & Bartley, N., (1985). Three-dimensional image processing using the concept of network resonance. *IEEE Transactions on Circuits and Systems, 32*(7), 664–672.
8. Coll, R. K., & Treagust, D. F., (2003). Investigation of secondary school, undergraduate, and graduate learners' mental models of ionic bonding. *Journal of Research in Science Teaching: The Official Journal of the National Association for Research in Science Teaching, 40*(5), 464–486.
9. Collman, J. P., & Boulatov, R., (2002). Heterodinuclear transition - metal complexes with multiple metal–metal bonds. *Angewandte Chemie International Edition, 41*(21), 3948–3961.
10. Cooper, M. M., Underwood, S. M., & Hilley, C. Z., (2012). Development and validation of the implicit information from Lewis structures instrument (IILSI): Do students connect structures with properties? *Chemistry Education Research and Practice, 13*(3), 195–200.
11. Curnow, O. J., & Pearson, J. K., (2023). Quantitative assessment on the effectiveness of a formal charge method for constructing Lewis

(electron dot) structures. *Journal of Chemical Education, 100(8), 3093-3097.*

12. de Oteyza, D. G., Gorman, P., Chen, Y. C., Wickenburg, S., Riss, A., Mowbray, D. J., & Fischer, F. R., (2013). Direct imaging of covalent bond structure in single-molecule chemical reactions. *Science, 340*(6139), 1434–1437.
13. Duffy, J. A., & Ingram, M. D., (1975). Optical basicity—IV: Influence of electronegativity on the Lewis basicity and solvent properties of molten oxyanion salts and glasses. *Journal of Inorganic and Nuclear Chemistry, 37*(5), 1203–1206.
14. Feinberg, M. J., & Ruedenberg, K., (1971). Paradoxical role of the kinetic - energy operator in the formation of the covalent bond. *The Journal of Chemical Physics, 54*(4), 1495–1511.
15. Filippetti, A., & Fiorentini, V., (2005). Coexistence of ionic and metallic bonding in noble-metal oxides. *Physical Review B, 72*(3), 035128.
16. Finkenstaedt-Quinn, S. A., Snyder-White, E. P., Connor, M. C., Gere, A. R., & Shultz, G. V., (2019). Characterizing peer review comments and revision from a writing-to-learn assignment focused on Lewis structures. *Journal of Chemical Education, 96*(2), 227–237.
17. Fukui, K., Yonezawa, T., Nagata, C., & Shingu, H., (1954). Molecular orbital theory of orientation in aromatic, heteroaromatic, and other conjugated molecules. *The Journal of Chemical Physics, 22*(8), 1433–1442.
18. Gade, L. H., (2000). Highly polar metal–metal bonds in "early–late" heterodimetallic complexes. *Angewandte Chemie International Edition, 39*(15), 2658–2678.
19. Gervasio, G., Bianchi, R., & Marabello, D., (2004). About the topological classification of the metal–metal bond. *Chemical Physics Letters, 387*(4–6), 481–484.
20. Goodman, R., Mannion, G., & Brzeski, A., (1996). *Reading Writing and Resonance: An Experiential Workshop for Practitioners, 1, 1-18.*
21. Gou, H., Hou, L., Zhang, J., & Gao, F., (2008). Pressure-induced incompressibility of ReC and effect of metallic bonding on its hardness. *Applied Physics Letters, 92*(24).
22. Gould, C. A., McClain, K. R., Reta, D., Kragskow, J. G., Marchiori, D. A., Lachman, E., & Long, J. R., (2022). Ultrahard magnetism from mixed-valence dilanthanide complexes with metal-metal bonding. *Science, 375*(6577), 198–202.

23. Habiddin, H., Farizka, L., & bin Shuid, A. N., (2023). Chemistry students' understanding of Lewis structure, VSEPR theory, molecular geometry, and symmetry: A Cross-sectional study. *JTK (Jurnal Tadris Kimiya), 8*(1), 1–9.
24. Hehre, W. J., Ditchfield, R., Radom, L., & Pople, J. A., (1970). Molecular orbital theory of the electronic structure of organic compounds. V: Molecular theory of bond separation. *Journal of the American Chemical Society, 92*(16), 4796–4801.
25. Hinze, J., & Jaffe, H. H., (1962). Electronegativity. I. Orbital electronegativity of neutral atoms. *Journal of the American Chemical Society, 84*(4), 540–546.
26. Huang, S., Kong, X., Xiong, Y., Zhang, X., Chen, H., Jiang, W., & Ren, C., (2020). An overview of dynamic covalent bonds in polymer material and their applications. *European Polymer Journal, 141*, 110094.
27. Kamogawa, K., Kaminaka, S., & Kitagawa, T., (1987). Behavior of ethanol in various binary solutions: difference Raman spectroscopy on the CH stretching vibrations. *Journal of Physical Chemistry, 91*(1), 222–226.
28. Karonen, M., Murtonen, M., Södervik, I., Manninen, M., & Salomäki, M., (2021). Heuristics hindering the development of understanding of molecular structures in university level chemistry education: *The Lewis Structure as an Example: Education Sciences, 11*(6), 258.
29. Kaya, S., Kaya, C., & Islam, N., (2016). Maximum hardness and minimum polarizability principles through lattice energies of ionic compounds. *Physica B: Condensed Matter, 485*, 60–66.
30. Keller, R. L. J., (2005). *Optimizing the Process of Nuclear Magnetic Resonance Spectrum Analysis and Computer Aided Resonance Assignment*. Doctoral dissertation, ETH Zurich, 1, 1-3.
31. Khanna, Z. S., Rao, B. K., & Jena, P., (2002). Electronic signature of the magicity and ionic bonding in Al 13 X (X= Li–K) clusters. *Physical Review B, 65*(12), 125105.
32. Khudhair, A. M., & Farid, M., (2021). A review on energy conservation in building applications with thermal storage by latent heat using phase change materials. *Thermal Energy Storage with Phase Change Materials*, 162–175.
33. Lever, A. B. P., (1972). Lewis structures and the octet rule. An automatic procedure for writing canonical forms. *Journal of Chemical Education, 49*(12), 819.

34. Levy Nahum, T., Mamlok - Naaman, R., Hofstein, A., & Taber, K. S., (2010). Teaching and learning the concept of chemical bonding. *Studies in Science Education, 46*(2), 179–207.
35. Longuet - Higgins, H. C., (1950). Some studies in molecular orbital theory I. Resonance structures and molecular orbitals in unsaturated hydrocarbons. *The Journal of Chemical Physics, 18*(3), 265–274.
36. Luxford, C. J., & Bretz, S. L., (2014). Development of the bonding representations inventory to identify student misconceptions about covalent and ionic bonding representations. *Journal of Chemical Education, 91*(3), 312–320.
37. Matskevich, N. I., Wolf, T., Greaves, C., Adelmann, P., Vyazovkin, I. V., & Matskevich, M. Y., (2015). New phases Bi12. 5Ln1. 5ReO24. 5: Thermodynamics and influence of dopant size on lattice energy (Ln: lanthanide). *The Journal of Chemical Thermodynamics, 91*, 234–239.
38. McArdle, P., (2019). Systematic procedure for drawing Lewis structures based on electron pairing priority and the explicit use of donor bonds: An alternative to the normal procedure which can be pen and paper based or automated on a PC in user interactive 3D. *Journal of Chemical Education, 96*(7), 1412–1417.
39. McGoran, E. C., (1991). Lone electron motion delocalization and relocalization to write Lewis structures. *Journal of Chemical Education, 68*(1), 19.
40. Miedema, A. R., De Boer, F. R., & De Chatel, P. F., (1973). Empirical description of the role of electronegativity in alloy formation. *Journal of Physics F: Metal Physics, 3*(8), 1558.
41. Moody, G. J., & Thomas, J. D. R., (1965). Lattice energy and chemical prediction: Use of the Kapustinskii equations and the Born-Haber cycle. *Journal of Chemical Education, 42*(4), 204.
42. Noorizadeh, S., & Shakerzadeh, E., (2008). A new scale of electronegativity based on electrophilicity index. *The Journal of Physical Chemistry A, 112*(15), 3486–3491.
43. O'Brien, M. H., & Akinc, M., (1990). Calculation of lattice energies and enthalpies of formation of rare-earth pyrosilicates. *Journal of Solid-State Chemistry, 84*(1), 130–137.
44. Özmen, H., (2004). Some student misconceptions in chemistry: A literature review of chemical bonding. *Journal of Science Education and Technology, 13*, 147–159.

45. Packer, J. E., & Woodgate, S. D., (1991). Lewis structures, formal charge, and oxidation numbers: A more user-friendly approach. *Journal of Chemical Education, 68*(6), 456.

46. Parr, R. G., & Pearson, R. G., (1983). Absolute hardness: Companion parameter to absolute electronegativity. *Journal of the American Chemical Society, 105*(26), 7512–7516.

47. Pearce, S. H., Williamson, C., Kifor, O., Bai, M., Coulthard, M. G., Davies, M., & Thakker, R. V., (1996). A familial syndrome of hypocalcemia with hypercalciuria due to mutations in the calcium-sensing receptor. *New England Journal of Medicine, 335*(15), 1115–1122.

48. Pearson, R. G., (1988). Absolute electronegativity and hardness: Application to inorganic chemistry. *Inorganic Chemistry, 27*(4), 734–740.

49. Petrov, D., (2014). Lattice enthalpies, polarizabilities, and shear moduli of lanthanide orthophosphates *LnPO 4. Acta Chim. Slov, 61*, 34–38.

50. Pople, J. A., Santry, D. P., & Segal, G. A., (1965). Approximate self - consistent molecular orbital theory. I: Invariant procedures. *The Journal of Chemical Physics, 43*(10), S129–S135.

51. Proniuk, S., & Blanchard, J., (2001). Influence of degree of substitution of cyclodextrins on their colligative properties in solution. *Journal of Pharmaceutical Sciences, 90*(8), 1086–1090.

52. Purser, G. H., (2001). Lewis structures in general chemistry: Agreement between electron density calculations and Lewis structures. *Journal of Chemical Education, 78*(7), 981.

53. Reddy, R. R., Ahammed, Y. N., Gopal, K. R., & Raghuram, D. V., (1998). Optical electronegativity and refractive index of materials. *Optical Materials, 10*(2), 95–100.

54. Reed, A. E., & Schleyer, P. V. R., (1990). Chemical bonding in hypervalent molecules. The dominance of ionic bonding and negative hyperconjugation over d-orbital participation. *Journal of the American Chemical Society, 112*(4), 1434–1445.

55. Reiher, M., & Sundermann, A., (2002). Do divalent [{HC (CR′ NR″)2}E] compounds contain E (I) or E (III) (E= B, Al, Ga, In)? on the correspondence of formal oxidation numbers, Lewis structures, and reactivity. *European Journal of Inorganic Chemistry, 2002*(7), 1854–1863.

56. Schomaker, V., & Stevenson, D. P., (1941). Some revisions of the covalent radii and the additivity rule for the lengths of partially ionic single covalent bonds. *Journal of the American Chemical Society, 63*(1), 37–40.
57. Schultz, N. E., Zhao, Y., & Truhlar, D. G., (2005). Databases for transition element bonding: Metal–metal bond energies and bond lengths and their use to test hybrid, hybrid meta, and meta density functionals and generalized gradient approximations. *The Journal of Physical Chemistry A, 109*(19), 4388–4403.
58. See, R. F., (2009). Which method of assigning bond orders in Lewis structures best reflects experimental data? An analysis of the octet rule and formal charge systems for period 2 and 3 nonmetallic compounds. *Journal of Chemical Education, 86*(10), 1241.
59. Seshadri, R., Felser, C., Thieme, K., & Tremel, W., (1998). Metal–metal bonding and metallic behavior in some ABO2 delafossites. *Chemistry of Materials, 10*(8), 2189–2196.
60. Shelley, J. C., Cholleti, A., Frye, L. L., Greenwood, J. R., Timlin, M. R., & Uchimaya, M., (2007). Epik: A software program for pK a prediction and protonation state generation for drug-like molecules. *Journal of Computer-Aided Molecular Design, 21*, 681–691.
61. Shultz, G. V., & Gere, A. R., (2015). Writing-to-learn the nature of science in the context of the Lewis dot structure model. *Journal of Chemical Education, 92*(8), 1325–1329.
62. Song, W., Martsinovich, N., Heckl, W. M., & Lackinger, M., (2013). Born–Haber cycle for monolayer self-assembly at the liquid–solid interface: Assessing the enthalpic driving force. *Journal of the American Chemical Society, 135*(39), 14854–14862.
63. Susen, S., (2020). The resonance of resonance: Critical theory as a sociology of world-relations? *International Journal of Politics, Culture, and Society, 33*(3), 309–344.
64. Suzuki, H., (2012). *Electronic Absorption Spectra and Geometry of Organic Molecules: An Application of Molecular Orbital Theory*. Elsevier, 1(1), 1-6.
65. Terada, Y., Ohkubo, K., Mohri, T., & Suzuki, T., (2002). Thermal conductivity of intermetallic compounds with metallic bonding. *Materials Transactions, 43*(12), 3167–3176.

66. Tichý, L., & Ticha, H., (1995). Covalent bond approach to the glass-transition temperature of chalcogenide glasses. *Journal of Non-Crystalline Solids, 189*(1–2), 141–146.

67. Weishaupt, D., Köchli, V. D., Marincek, B., Froehlich, J. M., Nanz, D., & Pruessmann, K. P., (2006). *How Does MRI Work?: An Introduction to the Physics and Function of Magnetic Resonance Imaging* (Vol. 2), pp. 3-10.

68. Xue, D., & Stains, M., (2020). Exploring students' understanding of resonance and its relationship to instruction. *Journal of Chemical Education, 97*(4), 894–902.

69. Yoder, C. H., & Flora, N. J., (2005). Geochemical applications of the simple salt approximation to the lattice energies of complex materials. *American Mineralogist, 90*(2–3), 488–496.

CHAPTER 6

The Effective Size of Atoms

Contents

6.1. INTRODUCTION

Within the expansive and complex field of atomic and molecular interactions, atomic size is a fundamental cornerstone, exerting significant influence over various chemical, physical, and biological phenomena. Comprehending the effective size of atoms holds significant importance in understanding matter's behavior at its fundamental level, encompassing various aspects such as chemical bond formation, material properties, and the operation of biological systems. The present chapter presents a thorough discussion of the dimensions of atoms, tracing their historical evolution and highlighting their importance in explaining the phenomena of the microscopic world (Smith et al., 2017).

The pursuit of comprehending the dimensions of atoms has a historical lineage spanning several centuries, coinciding with the unearthing of elements and chemical compounds. The conceptualization of atomic size in its early stages was inherently interconnected with the advancement of atomic theory (Ghanty & Ghosh, 1993; Sherby & Taleff, 2002).

The effective size of atoms heavily influences the determination of chemical reactivity and behavior of elements. The organization and interplay of atoms govern the establishment of chemical bonds, molecules' stability, and materials' characteristics. The atoms' dimensions significantly impact a range of observable properties at the macroscopic level, including the melting and boiling points of substances, their density, and their electrical and thermal conductivity (Kastner, 1993).

The reaction rates and the types of bonds that can be formed in chemical reactions are directly influenced by the effective size of atoms. A comprehensive detail of atomic size holds significant importance in predicting and manipulating chemical reactions, as well as in the development of novel molecules possessing tailored characteristics for diverse industrial applications, spanning from pharmaceuticals to materials science (Ghosh & Biswas, 2002).

During the early 20th century, the emergence of quantum mechanics led to the development of the concept of electron orbitals and their corresponding shapes. The Bohr model offered an initial insight into quantized energy levels and the arrangement of electron orbits encircling the atomic nucleus. Nevertheless, the application of quantum mechanics to the concept of atomic size significantly transformed our comprehension (Austin & Heine, 1966).

The utilization of Schrödinger's wave equation in quantum mechanics facilitated a more precise characterization of electron dynamics within atomic systems. Electron orbitals were introduced to depict regions characterized by a high probability density, indicating the likelihood of electron presence. The various orbitals, which possess distinct shapes and dimensions, played a crucial role in determining the effective atomic size and contributed to a more profound understanding of chemical bonding (Wang & Zunger, 1994).

The atomic size is a significant factor in chemical reactions, particularly bonding. The strength of chemical bonds is influenced by the interatomic distance, whereby shorter distances generally correspond to stronger bonds. This concept can be illustrated in covalent bonds, where the sharing of electrons between atoms occurs within a molecular orbital. The extent of overlap between these orbitals is contingent upon the respective atoms' sizes (Ghosh & Biswas, 2003).

Conversely, Ionic bonds originate from the electrostatic attraction between ions possessing opposite charges. The arrangement and packing in ionic compounds are influenced by the effective size of ions, resulting in the formation of diverse crystal structures (Lamb et al., 1980).

The atomic size heavily influences the determination of physical properties in materials. The arrangement of atoms within a material is influenced by their respective sizes, thereby exerting an impact on the substance's density and mechanical characteristics. For example, materials characterized by high atomic packing density exhibit greater density and hardness. In contrast, materials with a more open arrangement of atoms generally possess lower density and increased malleability (Osetsky et al., 2003). Moreover, the effective dimensions of atoms can potentially impact materials' electrical and thermal conductivity properties. The efficient electrical conduction in metals is attributed to free electrons, while the closely packed atoms facilitate heat transfer through the material.

Within the field of biology, the size of atoms plays a pivotal role in governing how biomolecules, such as enzymes and substrates, interact with one another. Enzymes possess active sites that exhibit complementary shapes and sizes, enabling them to interact with particular substrates, thereby facilitating biochemical reactions selectively. A thorough comprehension of the effective dimensions of atoms and the interactions between molecules is imperative to fully grasp the complexities of biochemical processes that govern biological life (Baji & Haeusler, 2022).

Based on the principles of wave mechanics, it can be observed that the electron density within an atom exhibits an asymptotic decrease toward zero as the distance from the atomic center increases. Hence, an atom lacks a precise magnitude. As two atoms approach one another, the strength of the interaction forces increases progressively (Cabot et al., 2001).

- One of the attractive forces consistently observed is the dispersion force, also known as the London attraction.
- Electronic interactions involve the formation of bonding molecular orbitals and the electrostatic attraction between atomic nuclei and electrons. These two factors contribute to the formation of bonding forces in covalent bonds.
- Electrostatic forces arise from the interactions between charged ions or atoms possessing partial charges of opposite signs.
- Repulsive interactions occur due to electrostatic forces between ions or partially charged atoms that possess charges of identical signs.
- Atomic nuclei experience electrostatic repulsion (Haken & Wolf, 2005).

The repulsion between electrons arises from their mutual electrostatic interactions and the repulsion resulting from the Pauli Exclusion Principle when electrons possess the same spin. The Pauli repulsion is the primary contributor to the repulsive forces.

This phenomenon arises from the fundamental principle that two electrons with identical spin orientations are prohibited from occupying the same spatial region. The phenomenon of Pauli repulsion can solely be discussed through the framework of quantum mechanics, thereby eluding straightforward conceptualization within simplistic models (Cao et al., 2021).

The efficacy of these forces exhibits variations, and their magnitudes vary to differing extents depending on the interatomic separation. The repulsion force mentioned previously exhibits superior effectiveness at proximity, albeit with limited range. At greater distances, the remaining forces exert a more prominent influence.

At a specific interatomic distance, there is an equilibrium between attractive and repulsive forces. The minimum point represents the equilibrium distance in a graph that depicts the relationship between potential energy and atomic distance (Giménez et al., 2006).

The concept of an equilibrium distance between atoms implies the notion of atoms as discrete entities with a discernible spherical shape. Indeed, it is often possible to consider atoms as approximately rigid spheres.

The assignment of multiple sphere radii is necessary for each atom due to the variation in attractive forces between atoms, which is contingent upon the specific type of bonding forces involved.

Based on empirical evidence, it has been observed that the atomic radius of an element remains relatively constant for a specific type of bonding. The following types of radii are distinguished: The radii that are commonly used in the field of chemistry include the Van Der Waals radii, metallic radii, various ionic radii that vary based on the specific ionic charges, as well as covalent radii that are associated with single, double, and triple bonds. In addition, it should be noted that the values exhibit variation based on the coordination numbers, whereby a higher coordination number corresponds to a larger radius (Zhang et al., 2013).

6.2. VAN DER WAALS RADII

In a crystalline compound composed of molecules, the molecules typically exhibit a close packing arrangement while ensuring that the atoms of adjacent molecules maintain a minimum distance equivalent to the combined values of their Van Der Waals radii.

Determining the Van Der Waals radius for an element involves measuring the shortest distance between atoms of the same element in adjacent molecules, which is commonly observed. A comprehensive examination discusses that atoms bonded through covalent interactions do not possess a perfectly spherical shape.

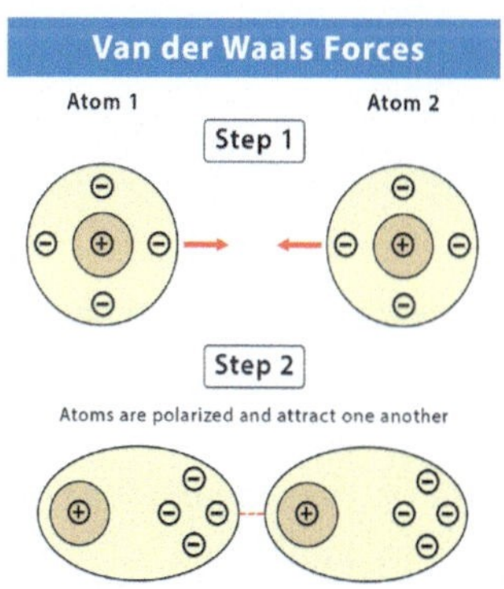

Figure 6.1. *Illustration of Van der Waals forces.*

Source: Chemistry Learner, Creative Commons License.

This phenomenon occurs when a halogen atom is bonded to a carbon atom. In this case, the halogen atom experiences a degree of flattening, resulting in a Van Der Waals radius that is shorter along the extension of the C-halogen bond than in perpendicular directions. In metal halides, when the covalent bond exhibits greater polarity, the extent of deviation from the spherical shape is reduced. The bonding type can also influence, as evidenced by the observation that carbon atoms in acetylenes possess a slightly larger atomic radius than carbon atoms in other compounds (Figure 6.1).

Instances where distances are shorter than the combined values of the Van Der Waals radii can arise due to the presence of unique, attractive forces. In the case of a solvated ion, determining distances between the ion and the atoms of the solvent molecules cannot be accomplished using Van Der Waals radii. The principle above holds when hydrogen bonding is present (Nag et al., 2007) (Figure 6.2).

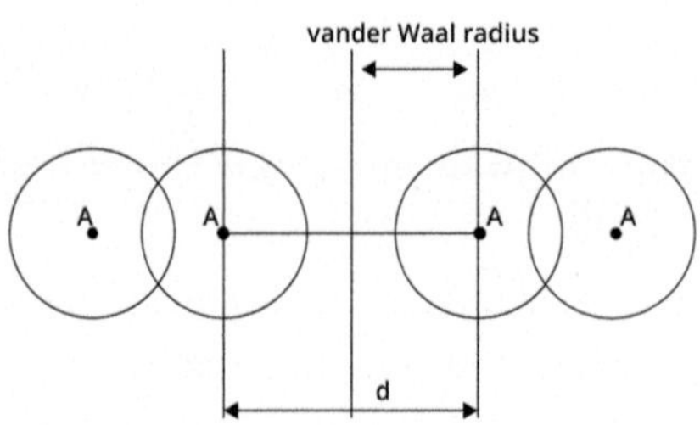

Figure 6.2. *Representation of Van der Waals radius.*

Source: Vedantu, Creative Commons License.

6.3. ATOMIC RADII

Atomic radii are essential characteristics that establish the spatial magnitude of an atom. The atomic radii correspond to the spatial separation between the atomic nucleus and the outermost electron shell, thereby determining the dimensions of the atom in its lowest energy state. Comprehending atomic radii is fundamental in analyzing diverse chemical and physical characteristics exhibited by elements and their compounds. This section explains the importance of atomic radii, the techniques employed to ascertain them, and their impact on the behavior of atoms and their interactions within chemistry and materials science (Nyburg & Faerman, 1985) (Figure 6.3).

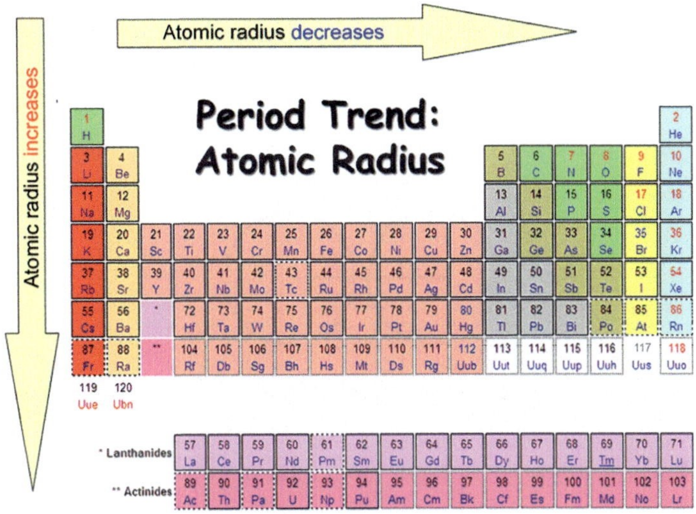

Figure 6.3. *Periodic trends for atomic radii.*

Source: Zach, Creative Commons License.

6.3.1. Significance of Atomic Radii

Determining atomic size is crucial in understanding the properties and behaviors of atoms, as atomic radii significantly impact these characteristics. The volume and arrangement of atoms in materials, as well as the density and other physical characteristics, are ultimately influenced by the radius of an atom. This concept illustrates that atoms possessing larger radii generally exhibit lower density due to their occupation of a greater volume relative to their mass.

Furthermore, determining atomic radii plays a pivotal role in comprehending chemical bonding and reactivity principles. The strength of chemical bonds is influenced by the interatomic distance within a molecule, which is determined by the cumulative atomic radii of the constituent atoms. Covalent bonds are formed when atoms share electrons through the overlap of their respective orbitals, and the atoms' atomic radii determine the bond's length. The arrangement and packing in the crystal lattice of ionic compounds are influenced by the size of ions, which is determined by their atomic radii (Figure 6.4).

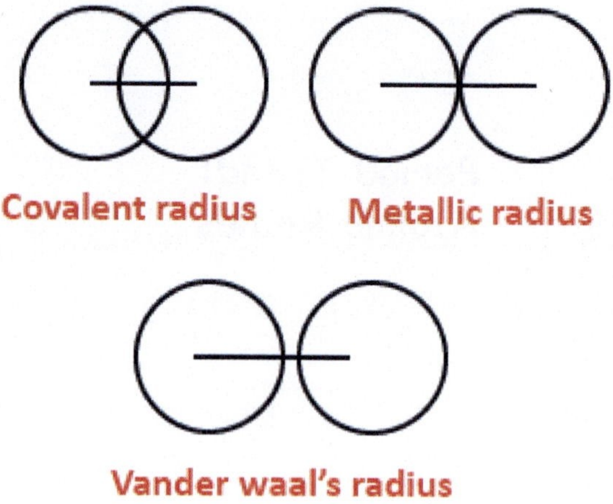

Figure 6.4. *Comparison of different atomic radii.*

Source: Byjus, Creative Commons License.

6.3.2. Methods for Determining Atomic Radii

Various experimental techniques are employed to ascertain atomic radii, each specifically designed to suit distinct contexts. X-ray crystallography is a widely recognized technique utilized to ascertain the spatial arrangement of atoms within a crystal lattice, thereby enabling the accurate determination of interatomic distances. Neutron diffraction is an additional robust methodology offering supplementary insights to X-ray crystallography, specifically in materials encompassing low-atomic mass elements.

Gas-phase electron diffraction is a technique that utilizes the interaction between electrons and atoms or molecules in the gaseous state to acquire structural data, such as bond lengths. Moreover, spectroscopic techniques, including optical and electron spectroscopy, offer valuable insights into the electronic configuration of atoms and the distribution of energy levels, which can be indirectly correlated with atomic radii (Chakraborty et al., 2010).

6.3.3. Trends in Atomic Radii

Systematic trends can be observed in the atomic radii across the periodic table. The atomic radii generally exhibit an upward trend as one progresses vertically within a group on the periodic table. The observed phenomenon can be ascribed to the progressive inclusion of electron shells characterized by higher principal quantum numbers (n), resulting in the enlargement of atomic orbitals and the expansion of the electron cloud.

In contrast, it is generally observed that atomic radii tend to decrease across a period horizontally, moving from left to right. The trend above can be attributed primarily to the augmentation of the effective nuclear charge, which arises from the progressive increment in the number of protons within the atomic nucleus. The enhanced nuclear attraction exerts a stronger force on the electrons, causing them to be drawn closer to the nucleus, thereby reducing the atomic radius.

6.3.4. Covalent and Van der Waals Radii

In molecules and materials, covalent radii and van der Waals radii hold significant importance in atomic radii. The covalent radii correspond to the mean separation between the nuclei of two atoms that are bonded together, predominantly within molecules connected by covalent bonds. The covalent radius is generally shorter than the van der Waals radius, primarily attributable to the substantial overlap of electron orbitals within covalent bonds.

In contrast, van der Waals radii pertain to measuring the separation between the nuclei of unbound atoms or molecules involved in van der Waals interactions. The radii above are linked to the relatively feeble intermolecular or interatomic forces arising from induced dipoles and London dispersion forces. The Van der Waals radii typically exhibit greater magnitudes than covalent radii, which can be attributed to the weaker interactions observed between non-bonded entities (Dvorak et al., 1992).

6.3.5. Applications in Chemistry and Materials Science

Understanding atomic radii is extensively utilized in diverse fields of study, such as chemistry and materials science. The formation and stability of molecules and compounds in chemical reactions are influenced by the size of atoms and their respective radii. A comprehensive detail of atomic radii is imperative to make accurate predictions and explain the reactivity and properties exhibited by chemical substances.

Determining the packing arrangements of atoms in crystalline structures in materials science heavily relies on considering atomic radii. The atoms' dimensions directly impact various material properties, such as lattice parameters, crystal density, and mechanical characteristics. The electronic properties of semiconductors can be significantly influenced by the dimensions of dopant atoms, thereby facilitating the customization of electronic devices. The concept of atomic radii is fundamental in comprehending the dimensions and characteristics of atoms. These principles govern numerous fundamental characteristics of elements and their compounds, exerting influence over chemical bonding, crystal structures, and the properties exhibited by materials. Determining atomic radii using experimental methods, such as X-ray crystallography and spectroscopy, yields significant insights into the atomic and sub-atomic domains, enhancing our comprehension of chemistry, materials science, and the natural environment. Examining atomic radius trends across the periodic table enhances our understanding of element behavior and provides a valuable means of predicting and interpreting chemical and physical phenomena. As scientific exploration of the microscopic realm progresses, the determination of atomic radii continues to play a fundamental role in enhancing our understanding of the complex concepts of atoms and molecules (Meyer & Dubois, 2013).

6.3.6. Atomic Radii in Metals

The cohesion of atoms in metals is determined by the relative dominance of bonding electron states over antibonding states in the electronic energy bands. Metals classified within the groups on the left side of the periodic table exhibit a limited number of valence electrons, resulting in a relatively low count of occupied bonding energy states. Elements on the right side of the periodic table exhibit a significant number of valence electrons, necessitating allocating a portion of these electrons into antibonding states. In both instances, there exists a presence of comparatively feeble metallic bonding. When many bonding states are occupied, and only a few antibonding states are occupied, the resultant interatomic forces between the metal atoms are substantial. This statement holds for the transition metals located within the central region of the periodic table. The atomic radii of metals exhibit a decreasing trend from the alkali metals to the metals from groups six to eight, followed by an increasing trend. The general trend of decreasing atomic sizes observed across all periods from the alkali metals to the noble gases is superimposed on this sequence. This trend can be attributed to the increasing nuclear charge. The valence electron concentration, which refers to the ratio of the total number of available valence electrons to the number of atoms, plays a crucial role in determining the effective atomic size of intermetallic compounds (Biryukov & Struchkov, 1970) (Figure 6.5).

Li 152	Be 112												
Na 186	Mg 160											Al 143	
K 230	Ca 197	Sc 162	Ti 146	V 134	Cr 128	Mn 137	Fe 126	Co 125	Ni 125	Cu 128	Zn 134	Ga 135	
Rb 247	Sr 215	Y 180	Zr 160	Nb 146	Mo 139	Tc 135	Ru 134	Rh 134	Pd 137	Ag 144	Cd 151	In 167	Sn 154
Cs 267	Ba 222	La 187	Hf 158	Ta 146	W 139	Re 137	Os 135	Ir 136	Pt 139	Au 144	Hg 151	Tl 171	Pb 175

Ce 182	Pr 182	Nd 182	Pm 181	Sm 180	Eu 204	Gd 179	Tb 178	Dy 177	Ho 176	Er 175	Tm 174	Yb 193	Lu 174
Th 180	Pa 161	U 156	Np 155	Pu 159	Am 173	Cm 174	Bk 170	Cf 169	Es	Fm	Md	No	Lr

Figure 6.5. *Atomic radii (pm) in metals. All values refer to a coordination number 12, except for the alkali metals.*

Source: Muller, Creative Commons License.

Did You Know?

The length of covalent bonds can vary significantly between different elements and molecules. For example, the carbon-carbon bond in diamond is one of the shortest, while the carbon-carbon single bond in ethane is longer.

6.4. COVALENT RADII

Determining covalent radii is of utmost importance in chemistry as it is a fundamental parameter in explaining the nature of chemical bonding and molecular geometry. The values provided denote the atomic radii of atoms that are interconnected through a covalent bond, wherein electrons are shared between said atoms. The emergence of the concept of covalent radii was driven by the necessity to provide an accurate description of interatomic distances within molecules, thereby explaining the characteristics of chemical bonds and molecular properties (Table 6.1).

Table 6.1. Covalent Radii of the Halogen Group Elements

Atom	Covalent Radius (pm)	Nuclear Charge
F	64	+9
Cl	99	+17
Br	114	+35
I	133	+53
At	148	+85

The covalent radii can be defined as the measure of half the distance separating the nuclei of two atoms that are joined together using a covalent bond. The radii above serve as a quantitative indicator of the relative dimensions of atoms during their participation in the chemical bonding process. In contrast to atomic radii, covalent radii consider the impact of neighboring atoms on the electron cloud, thereby reflecting the collaborative characteristic of electrons in a covalent bond.

The determination of covalent radii can be achieved through a range of experimental and theoretical approaches encompassing diverse methodologies (Schomaker & Stevenson, 1941):

1. **X-Ray Crystallography:** By examining X-ray diffraction patterns of crystalline compounds, researchers can derive the interatomic distances within the crystal lattice. These distances are indicative of the covalent radii present in covalent bonds.
2. **Molecular Modeling:** Computational techniques, such as quantum mechanical computations and molecular dynamics simulations, can offer theoretical approximations of covalent radii by leveraging the electronic configuration of molecules.
3. **Bond Lengths in Molecules:** The acquisition of experimental measurements about bond lengths in molecules is important in determining covalent radii (Politzer et al., 2003).

Similar to atomic radii, covalent radii exhibit trends across the periodic table:

1. **Periodic Trend:** The trend observed in covalent radii is a general decrease as one moves from left to right across a period. As one moves from left to right across the periodic table, the size of atoms decreases. Consequently, the shared electrons in chemical bonds are held in closer proximity to the nuclei, leading to a reduction in bond lengths.
2. **Group Trend:** The trend observed in covalent radii is an increase in size as one moves down a group. As an element descends within a group, there is an increase in the number of electron shells, resulting in an expansion of atomic size and elongation of covalent bond lengths.

The concept of covalent radii is fundamental in understanding the nature of chemical bonds:

1. **Bond Lengths:** The length of covalent bonds in molecules is directly influenced by covalent radii. Stronger bonds are characterized by

shorter bond lengths, which can be attributed to the nuclei being in closer proximity. This proximity results in higher electron density and a more pronounced overlap of atomic orbitals.

2. **Bond Strengths:** The strength of covalent bonds is determined by the extent of atomic orbital overlap between bonded atoms, which is influenced by their respective covalent radii. Greater degrees of overlap between atomic orbitals lead to the formation of more robust chemical bonds characterized by elevated bond dissociation energies.
3. **Bond Angles and Molecular Geometry:** The three-dimensional arrangement of atoms in molecules is influenced by covalent radii. The molecular geometry and shapes of molecules are defined by the spatial distribution of atoms, which is determined by bond lengths and angles.

Covalent radii refer to measuring the distances between atoms participating in covalent bonding. Conversely, van der Waals radii denote the effective dimensions of atoms engaged in non-bonded interactions. In general, the covalent radii of atoms tend to be smaller than their van der Waals radii. This discrepancy arises because atoms are more densely arranged when they form chemical bonds within molecules, unlike their non-bonded interactions (Gillespie & Johnson, 1997). The significance of covalent radii is of utmost importance in diverse domains of chemistry and materials science:

1. **Chemical Reactivity:** The comprehension of covalent radii facilitates the anticipation of molecular reactivity and the probability of stable compound formation.
2. **Molecular Design:** Using covalent radii facilitates molecular design, enabling scientists to engineer molecules with desired properties and functionalities strategically.
3. **Materials Engineering:** The crystal structures and properties of materials are influenced by covalent radii, which guide the design of novel materials with specific desired characteristics.

Covalent radii play a crucial role in contemporary chemistry as they provide valuable insights into the characteristics of chemical bonds, molecular geometry, and the behavior of atoms within molecules. The research conducted by the authors has significantly transformed the comprehension of the composition and characteristics of matter, establishing the fundamental basis for the emergence of novel materials and advancements across multiple scientific and technological domains. Ongoing investigations into covalent radii can unveil novel findings in chemistry and materials science, thereby influencing the trajectory of scientific inquiry and advancement.

Determining covalent radii involves analyzing the measured interatomic distances between atoms that are covalently bonded to one another and belong to the same chemical element.

This can be illustrated by comparing the C–C bond length between diamond and alkanes, which measures 154 pm. It is worth noting that this value represents the covalent radius for a single bond at a carbon atom with a coordination number of 4 (sp_3 C atom), with half of this value being 77 pm.

The covalent radii for chlorine (100 pm), oxygen (73 pm), and silicon (118 pm) can be determined by measuring the respective distances in their corresponding molecules, namely Cl_2, O_2, and elemental silicon. By summing the covalent radii of carbon (C) and chlorine (Cl), a value of 77 + 100 = 177 is obtained. This value aligns reasonably well with the observed distances in compounds containing a carbon-chlorine (C-Cl) bond (Hahn & Trebin, 2000).

Nevertheless, when the covalent radii of silicon (Si) and oxygen (O) are combined, resulting in a sum of 118 pm for Si and 73 pm for O, the obtained value of 191 pm does not align adequately with the observed distances in silicon dioxide (SiO_2), which range from 158 to 162 pm. In general, it is necessary to assert that as the polarity of a bond increases, its length exhibits a greater deviation towards lower values of the sum of the covalent radii. To consider this aspect, SHOMAKER, and STEVENSON derived the subsequent correction formula:

$$d\,(AX) = r\,(A) + r\,(X) - c|X\,(A) - x\,(X)|$$

where; d(AX) is the bond length, r(A) und r(X) is the covalent radii of the atoms A and X, x(A); and x(X) is the electronegativities of A and X (Li et al., 2022).

The correction parameter (c) is contingent upon the specific atoms under consideration and exhibits values ranging from 2 to 9 picometers (pm). No correction is required for C–X bonds, except N, O, and F, when X belongs to the 5th, 6th, or 7th main groups. The manifestation of bond polarity is evident in the correlation between bond lengths and oxidation states. For instance, the P–O bonds in P_4O_6 exhibit a greater length of 164 pm than the P_4O_{10} bonds, which measure 160 pm (with a sum of covalent radii of

183 pm). The magnitudes of these deviations are greater for atoms with higher polarizability, commonly called 'soft' atoms. The determination of bond lengths can be subject to uncertainty in cases where the degree of multiple bonding is not precisely known, the influence of lone electron pairs on adjacent bonds is unclear, and the extent to which the ionic charge and coordination number affect the bond lengths is uncertain. The variation in Cl–O bond lengths serves as a demonstration of this phenomenon. The bond lengths of the following compounds are as follows: HOCl at 170 pm, ClO_2 at 156 pm, ClO_3 at 149 pm, ClO_4 at 143 pm, $HClO_3$ with one bond at 164 pm, and three bonds at 141 pm, ClO_2 at 147 pm, and ClO_2^+ at 131 pm (Su et al., 2019).

6.5. IONIC RADII

Ionic radii play a pivotal role in determining the dimensions of ions, which are electrically charged entities resulting from atoms' gain or loss of electrons. Examining ionic radii yields significant knowledge regarding the ionic characteristics of compounds, the organization of ions within crystal structures, and the involvement of ions in chemical and biological systems. This investigation into the concept of ionic radii examines their definition, determination, and importance across different scientific fields, uncovering their fundamental contribution to comprehending the behavior of ions and their interactions (Shannon & Prewitt, 1970) (Figure 6.6).

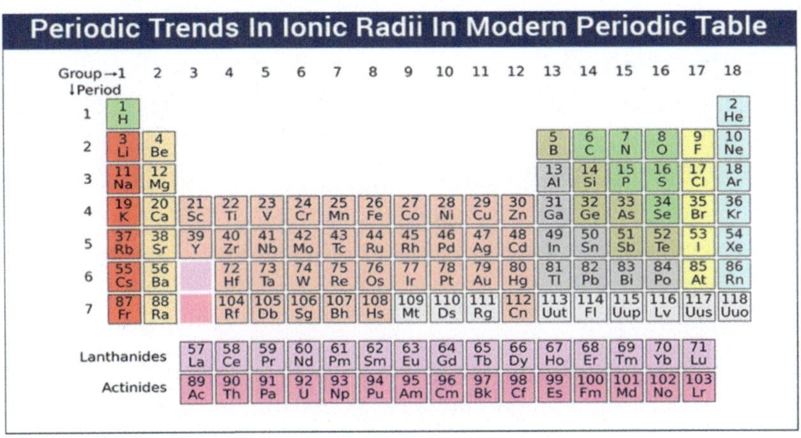

Figure 6.6. *Periodic trends in ionic radii (pm) in the modern periodic table.*

Source: Nicolas, Creative Commons License.

Ionic radii pertain to the dimensions of ions, which arise from electron transfer between atoms. When an atom undergoes electron loss, resulting in a positive charge (cation), its size diminishes due to the diminished electron cloud. On the other hand, if an atom acquires electrons resulting in a negative charge (anion), its size expands due to the augmented electron cloud.

Determining ionic radii through experimental means presents challenges owing to the inherent dynamism of ions and their active participation in chemical reactions. There are several methodologies utilized in the determination of ionic radii, which encompass the following (D'Angelo et al., 2011):

1. **X-Ray Crystallography:** X-ray diffraction analyses conducted on crystalline forms of ionic compounds yield valuable insights into the interionic distances within the crystal lattice, specifically between cations and anions.
2. **Coulomb's Law:** The interplay of attractive and repulsive forces among ions within a crystal lattice can be effectively described by applying Coulomb's law. This theoretical framework offers valuable insights into the determination of ionic sizes.
3. **Pauling's Rules:** The formulation of rules by Linus Pauling pertains to the description of the correlation between the dimensions of ions and the coordination number, which denotes the number of neighboring ions within a crystal lattice.

Ionic radii exhibit trends across the periodic table:

1. **Periodic Trend:** The trend observed in the periodic table indicates a general decrease in ionic radii as one moves from the left to the right within a given period. The observed phenomenon can be attributed to the progressive augmentation of the effective nuclear charge, leading to a heightened electrostatic attraction between the electrons and the nucleus. Consequently, this interaction results in the formation of ions with reduced dimensions (Deblonde et al., 2021).
2. **Group Trend:** The trend observed in groups of elements is that the ionic radii tend to increase as one moves down the group. The progressive increase in electron shells within the group results in the formation of larger ions characterized by expanded electron cloud dimensions.

The significance of ionic radii is of utmost importance in numerous scientific fields:

1. **Crystal Structures:** The ionic radii significantly influence the determination of the ions' arrangement in crystal structures. The dimensions of cations and anions play a significant role in determining the arrangement and structure of the crystal lattice, thereby impacting the characteristics and behaviors of ionic compounds.
2. **Ionic Bonding:** The size of ions directly influences the strength of ionic bonds. Larger ions generally form weaker bonds due to decreased electrostatic attraction, while smaller ions form stronger bonds.
3. **Solubility and Precipitation:** Ionic radii influence the solubility and precipitation of ionic compounds. In solutions, ions with similar sizes are more likely to be compatible and dissolve effectively.
4. **Biological Systems:** In biological systems, ionic radii interact with ions and biomolecules, including ion channels, enzymes, and transporters.

In coordination chemistry, ionic radii are essential in understanding the formation of coordination complexes. The size of the metal ion determines the geometry and stability of the complex, influencing its reactivity and properties. Ionic radii are essential parameters that provide insights into the sizes of ions and their behavior in chemical and biological systems. Understanding ionic radii is central to comprehending ionic bonding, crystal structures, and the behavior of ions in solutions and complex environments. As the knowledge of ionic radii advances, it continues to shape the understanding of the properties and interactions of ions, paving the way for advancements in various fields of chemistry, materials science, and biology (Morgan et al., 2018).

The shortestcation–anion distance in an ionic compound corresponds to the sum of the ionic radii. This distance can be determined experimentally. However, there is no straightforward way to obtain values for the radii themselves. Data taken from carefully performed X-ray diffraction experiments allow the crystal's electron density calculation; the point having the minimum electron density along the connection line between a cation and an adjacent anion can be taken as the contact point of the ions. As shown in the example of sodium fluoride in Figure 6.7, the ions in the crystal show certain deviations from spherical shape, i.e., the electron shell is polarized. This indicates the presence of some degree of covalent bonding, which can be interpreted as a partial backflow of electron density from the anion to the cation. The electron density minimum, therefore, does not necessarily represent the ideal place for the limit between cation and anion.

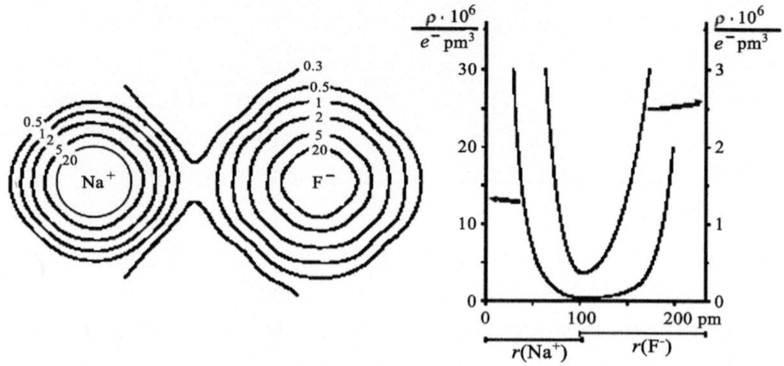

Figure 6.7. *Experimentally determined electron density (ρ) in crystalline sodium fluoride.*

Source: Catherine, Creative Commons License.

The commonly used values for ionic radii are based on an arbitrarily assigned standard radius for a certain ion. This way, a consistent set of radii for other ions can be derived. Several tables have been published, each derived using a different standard value for one ionic radius (ionic radii by Goldschmidt, Pauling, Ahrens, & Shannon). The values by Shannon are based on a critical evaluation of experimentally determined interatomic distances and the standard radius of 140 pm for the O_2 ions with six-fold coordination.

Ionic radii can also be used when considerable covalent bonding is involved. The higher the charge of a cation, the greater its polarizing effect on a neighboring anion, i.e., the covalent character of the bond increases. Nevertheless, arithmetically, one can still assume a constant radius for the anion and assign a radius for the cation that will yield the correct interatomic distance. A value like r(Nb^{5+}) = 64 pm, therefore, does not imply the existence of an Nb^{5+} ion with this radius but means that in niobium (V) compounds, the bond length between an Nb atom and a more electronegative atom X can be calculated as the sum of r(Nb^{5+}) plus the anionic radius of X. However, the values are not completely independent of the nature of X; for this purpose, another set of slightly different ionic radii has been derived. Conversely, one can deduce the oxidation states of the atoms from observed bond lengths (Tansel et al., 2006).

Ionic radii of soft (easily polarized) ions depend on the counter-ion. The H ion is an example; its radius is 130 pm in MgH_2, 137 pm in LiH, 146 pm in NaH, and 152 pm in KH.

H -1 ~150	Li +1 76	Be +2 45	B +3 27	C +4 16	N -3 146 +3 16	O -2 140	F -1 133
	Na +1 102	Mg +2 72	Al +3 54	Si +4 40	P +3 44 +5 38	S -2 184 +6 29	Cl -1 181
	K +1 138	Ca +2 100	Ga +3 62	Ge +2 73 +4 53	As +3 58 +5 46	Se -2 198 +4 50	Br -1 196
	Rb +1 152	Sr +2 118	In +3 80	Sn +2 118 +4 69	Sb +3 76 +5 60	Te -2 221 +4 97 +6 56	I -1 220 +5 95 +7 53
	Cs +1 167	Ba +2 135	Tl +1 150 +3 89	Pb +2 119 +4 78	Bi +3 103 +5 76	Po +4 94 +6 67	

Figure 6.8. *Ionic radii (pm) for main group elements.*

Source: Catherine, Creative Commons License.

In most cases, the ionic radii in Figure 6.8 apply to ions with coordination number 6. For other coordination numbers, slightly different values must be taken. For every unit by which the coordination number increases or decreases, the ionic radius increases or decreases by 1.5 to 2%. For coordination number 4, the values are approximately 4% smaller than for 6, and for coordination number 8, about 3% greater than for coordination number 6. The reason for this is the mutual repulsion of the coordinated ions (El-Sonbati et al., 2015).

6.6. CLASSICAL PICTURE OF ATOMIC SIZE

The classical picture of atomic size represents the early understanding of atoms and their structure before the advent of quantum mechanics. It emerged from the efforts of early scientists to unravel the nature of matter and laid the groundwork for the development of modern atomic theory. This section explores the historical journey of atomic investigations, from ancient philosophical speculations to the atomic models proposed by pioneering scientists in the 19th and early 20th centuries. While the classical picture provided a starting point, it also revealed its limitations, paving the way for the revolutionary quantum mechanical treatment of atomic size that followed.

6.6.1. The Ancient Philosophers and Indivisible Particles

The concept of atoms as indivisible particles was first proposed by the ancient Greek philosophers, notably Democritus and Leucippus, around the 5th century BCE. They postulated that all matter was composed of tiny, indestructible units called "atoms," derived from the Greek word "atoms," meaning indivisible. This philosophical idea marked the earliest inklings of atomic theory and laid the foundation for later scientific investigations into the microscopic world of atoms and their size (Newman, 1996).

6.6.2. The Daltonian Atom

In Dalton's model, atoms were envisioned as solid, sphere-like particles, all of the same size and mass for a given element. The idea of atoms as building blocks of matter was transformative, setting the stage for further investigations into the nature of atoms and their effective size (Grossman, 2014).

Fun Fact

Due to the wave nature of electrons, there is a probability that an electron can "tunnel" through a barrier that it classically does not have enough energy to overcome. This phenomenon is essential in technologies like scanning tunneling microscopes (STMs) used to image individual atoms.

6.6.3. The Plum Pudding Model

Thomson's model incorporated the discovery of the electron as a subatomic particle, providing valuable insights into atomic structure (Lynch & Dawson, 2004). However, the plum pudding model could not fully explain certain experimental results, such as the scattering of alpha particles in the famous gold foil experiment conducted by Ernest Rutherford in 1911. This pivotal experiment led to the unraveling of the true nature of atomic size and the presence of a dense, positively charged nucleus at the center of the atom (Liu et al., 2020) (Figure 6.9).

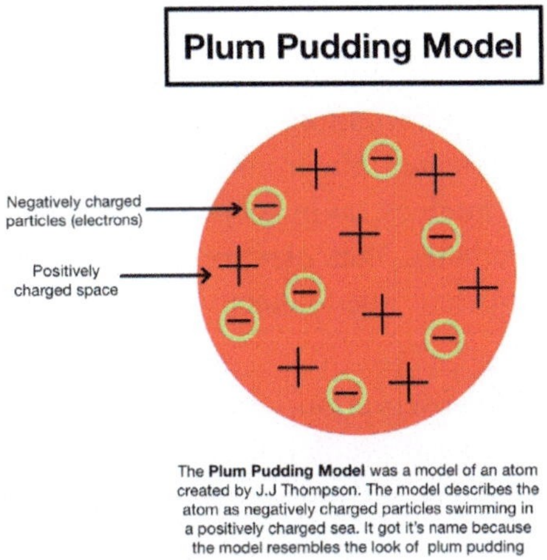

Figure 6.9. *The plum pudding model of an atom.*

Source: Herald, Creative Commons License.

6.6.4. The Rutherford Nuclear Model

The Rutherford model introduced the concept of nuclear size, with most of the atom's mass concentrated in the nucleus and the electrons occupying a much larger volume around it. This understanding paved the way for further investigations into the arrangement and behavior of electrons within the atom (Jackman et al., 1982; Rauscher et al., 2013) (Figure 6.10).

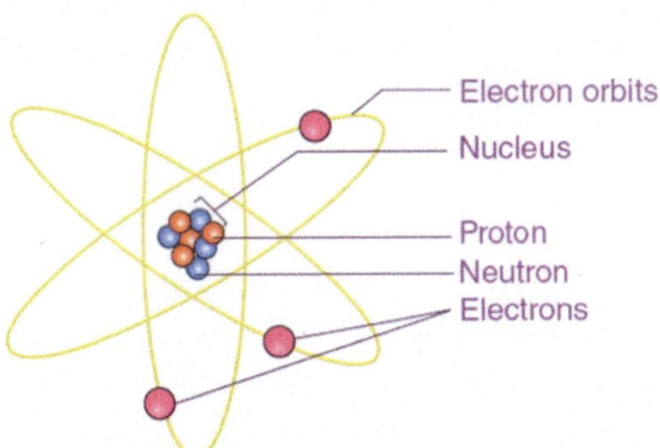

Figure 6.10. *The Rutherford nuclear model of atom.*

Source: Nicholas, Creative Commons License.

6.6.5. Limitations of Classical Models

While the classical picture of atomic size provided a foundation for studying atoms, it had inherent limitations. The models proposed by Dalton, Thomson, and Rutherford treated electrons as particles moving in well-defined orbits around the nucleus, akin to planets orbiting the sun. This classical interpretation predicted that electrons should continuously lose energy and spiral into the nucleus due to electromagnetic radiation (Johnson et al., 2017).

However, experimental observations, particularly from spectroscopy, did not align with these classical predictions. The study of atomic spectra revealed discrete, quantized energy levels, leading to the realization that electrons did not move in classical orbits but occupied specific, quantized energy states. This discrepancy signaled the breakdown of classical physics at the atomic scale and set the stage for revolutionary developments in quantum mechanics (Rusch et al., 2017).The classical picture of atomic size, rooted in ancient philosophical ponderings and the early atomic theories of Dalton, Thomson, and Rutherford, provided an essential starting point for understanding atoms and their properties. These classical models laid the groundwork for the development of modern atomic theory and stimulated further investigations into the microscopic world (Jagiello et al., 2016). However, the limitations of classical models became apparent as experimental evidence accumulated, revealing classical physics' inadequacy in accurately describing atomic behavior. The true nature of atomic size and the behavior of electrons within atoms required a paradigm shift, leading to the emergence of quantum mechanics and the quantum mechanical treatment of atomic size, revolutionizing our understanding of the atomic structure (Leonov et al., 2015).

6.7. SUMMARY

The chapter on the effective size of atoms explores the fascinating world of atomic dimensions, providing valuable insights into the behavior and interactions of matter at its most fundamental level. It illustrates several distinct radii that define atoms' size and significance in various scientific contexts. Van der Waals radii, atomic radii, covalent radii, and ionic radii represent effective atomic size. Van der Waals radii describe the effective size of atoms and molecules in non-bonded interactions, influencing the packing of atoms in solids and non-covalent interactions in gases, liquids, and solids. Atomic radii define the distance between the nucleus and the outermost shell, influencing chemical bonding, molecular properties, and the arrangement of atoms in the periodic table. Covalent radii represent the sizes of atoms involved in chemical bonding, which is crucial in determining the strength and nature of covalent bonds. They are significant in molecular design and materials engineering. Ionic radii define the sizes of ions formed when atoms gain or lose electrons, influencing crystal structures, ionic bonding, and the behavior of ionic compounds. The classical understanding of atomic size, derived from early philosophical speculations and the work of Dalton, Thomson, and Rutherford, laid the foundation for atomic exploration. However, the development of quantum mechanics revolutionized the understanding of atomic behavior, introducing electron orbitals and quantized energy levels, which provide a more accurate depiction of atomic size. In conclusion, the chapter on the effective size of atoms uncovers the importance of atomic dimensions in various scientific disciplines. Effective atomic size influences molecular interactions, chemical bonding, and material properties. The historical context of classical atomic models highlights the transformative impact of quantum mechanics on the understanding of atomic behavior. As research in this field advances, it promises to deepen the understanding of the microscopic universe and pave the way for discoveries and applications in science and technology.

REFERENCES

1. Austin, B. J., & Heine, V., (1966). Pseudopotentials, the sizes of atoms and their s—p Splittings. *The Journal of Chemical Physics*, *45*(3), 928–933.
2. Baji, F., & Haeusler, C., (2022). Introducing Iranian primary children to atoms and molecules. *Research in Science Education*, *52*(4), 1387–1418.
3. Biryukov, B. P., & Struchkov, Y. T., (1970). Metal–metal bonds and covalent atomic radii of transition metals in their π-complexes and polynuclear carbonyls. *Russian Chemical Reviews*, *39*(9), 789.
4. Cabot, A., Dieguez, A., Romano-Rodrıguez, A., Morante, J. R., & Barsan, N., (2001). Influence of the catalytic introduction procedure on the nano-SnO_2 gas sensor performances: Where and how stay the catalytic atoms? *Sensors and Actuators B: Chemical*, *79*(2–3), 98–106.
5. Cao, Y., Guo, L., Dan, M., Doronkin, D. E., Han, C., Rao, Z., & Zhou, Y., (2021). Modulating electron density of vacancy site by single Au atom for effective CO_2 photoreduction. *Nature Communications*, *12*(1), 1675.
6. Chakraborty, T., Gazi, K., & Ghosh, D. C., (2010). Computation of the atomic radii through the conjoint action of the effective nuclear charge and the ionization energy. *Molecular Physics*, *108*(16), 2081–2092.
7. D'Angelo, P., Zitolo, A., Migliorati, V., Chillemi, G., Duvail, M., Vitorge, P., & Spezia, R., (2011). Revised ionic radii of lanthanoid (III) ions in aqueous solution. *Inorganic Chemistry*, *50*(10), 4572–4579.
8. Deblonde, G. J. P., Zavarin, M., & Kersting, A. B., (2021). The coordination properties and ionic radius of actinium: A 120-year-old enigma. *Coordination Chemistry Reviews*, *446*, 214130.
9. Dvorak, M. A., Ford, R. S., Suenram, R. D., Lovas, F. J., & Leopold, K. R., (1992). van der Waals vs. covalent bonding: Microwave characterization of a structurally intermediate case. *Journal of the American Chemical Society*, *114*(1), 108–115.
10. El-Sonbati, A. Z., Diab, M. A., El-Bindary, A. A., Eldesoky, A. M., & Morgan, S. M., (2015). Correlation between ionic radii of metals and thermal decomposition of supramolecular structure of azodye complexes. *Spectrochimica Acta Part A: Molecular and Biomolecular Spectroscopy*, *135*, 774–791.
11. Ghanty, T. K., & Ghosh, S. K., (1993). Correlation between hardness,

polarizability, and size of atoms, molecules, and clusters. *The Journal of Physical Chemistry*, *97*(19), 4951–4953.

12. Ghosh, D. C., & Biswas, R., (2002). Theoretical calculation of absolute radii of atoms and ions. Part 1. The atomic radii. *International Journal of Molecular Sciences*, *3*(2), 87–113.
13. Ghosh, D. C., & Biswas, R., (2003). Theoretical calculation of absolute radii of atoms and ions. Part 2. The ionic radii. *International Journal of Molecular Sciences*, *4*(6), 379–407.
14. Gillespie, R. J., & Johnson, S. A., (1997). Study of bond angles and bond lengths in disiloxane and related molecules in terms of the topology of the electron density and its Laplacian. *Inorganic Chemistry*, *36*(14), 3031–3039.
15. Giménez, D., Andreu, C., Varea, T., Diaz, D., & Asensio, G., (2006). The introduction of fluorine atoms or trifluoromethyl groups in short cationic peptides enhances their antimicrobial activity. *Bioorganic & Medicinal Chemistry*, *14*(20), 6971–6978.
16. Grossman, M. I., (2014). John Dalton and the London atomists: William and Bryan Higgins, William Austin, and new Daltonian doubts about the origin of the atomic theory. *Notes and Records: the Royal Society Journal of the History of Science*, *68*(4), 339–356.
17. Hahn, J., & Trebin, H. R., (2000). Molecular dynamics of covalent crystals. In: *High Performance Computing in Science and Engineering'99: Transactions of the High Performance Computing Center Stuttgart (HLRS)* (Vol. 2, pp. 92–99). Berlin, Heidelberg: Springer Berlin Heidelberg–1999.
18. Haken, H., & Wolf, H. C., (2005). *The Physics of Atoms and Quanta: Introduction to Experiments and Theory* (Vol. 1439, pp. 34–38). Springer Science & Business Media.
19. Jackman, T. E., Davies, J. A., Jackson, O. P., Unertl, W. N., & Norton, P. R., (1982). The Pt (110) phase transitions: A study by Rutherford backscattering, nuclear microanalysis, LEED, and thermal desorption spectroscopy. *Surface Science*, *120*(2), 389–412.
20. Jagiello, K., Grzonkowska, M., Swirog, M., Ahmed, L., Rasulev, B., Avramopoulos, A., & Puzyn, T., (2016). Advantages and limitations of classic and 3D QSAR approaches in nano-QSAR studies based on biological activity of fullerene derivatives. *Journal of Nanoparticle Research*, *18*, 1–16.

21. Johnson, J. B., Duvoy, P. X., Kulchitsky, A. V., Creager, C., & Moore, J., (2017). Analysis of mars exploration rover wheel mobility processes and the limitations of classical terramechanics models using discrete element method simulations. *Journal of Terramechanics*, *73*, 61–71.

22. Kastner, M. A., (1993). Artificial atoms. *Physics Today*, *46*(1), 24–31.

23. Lamb, J. D., Izatt, R. M., Swain, C. S., & Christensen, J. J., (1980). A systematic study of the effect of macrocycle ring size and donor atom type on the log K. Delta. H, and T. Delta. S of reactions at 25°C in methanol of mono-and divalent cations with crown ethers. *Journal of the American Chemical Society*, *102*(2), 475–479.

24. Leonov, G. A., Kuznetsov, N. V., Yuldashev, M. V., & Yuldashev, R. V., (2015). Hold-in, pull-in, and lock-in ranges of PLL circuits: rigorous mathematical definitions and limitations of classical theory. *IEEE Transactions on Circuits and Systems I: Regular Papers*, *62*(10), 2454–2464.

25. Li, S., Liu, Y., Chen, D., Jiang, Y., Nie, Z., & Pan, F., (2022). Encoding the atomic structure for machine learning in materials science. *Wiley Interdisciplinary Reviews: Computational Molecular Science*, *12*(1), 1558.

26. Liu, Z., Wang, J., & Lu, B., (2020). Plum pudding model inspired KVPO4F@3DC as high-voltage and hyperstable cathode for potassium ion batteries. *Science Bulletin*, *65*(15), 1242–1251.

27. Lynch, I., & Dawson, K. A., (2004). Release of model compounds from "plum-pudding"-type gels composed of microgel particles randomly dispersed in a gel matrix. *The Journal of Physical Chemistry B*, *108*(30), 10893–10898.

28. Meyer, F., & Dubois, P., (2013). Halogen bonding at work: recent applications in synthetic chemistry and materials science. *CrystEngComm*, *15*(16), 3058–3071.

29. Morgan, S. M., El-Ghamaz, N. A., & Diab, M. A., (2018). Effect of the type of metal on the electrical conductivity and thermal properties of metal complexes: The relation between ionic radius of metal complexes and electrical conductivity. *Journal of Molecular Structure*, *1160*, 227–241.

30. Nag, S., Banerjee, K., & Datta, D., (2007). Estimation of the van der Waals radii of the d-block elements using the concept of bond valence. *New Journal of Chemistry*, *31*(6), 832–834.

31. Newman, W. R., (1996). The alchemical sources of Robert Boyle's corpuscular philosophy. *Annals of Science*, *53*(6), 567–585.
32. Nyburg, S. C., & Faerman, C. H., (1985). A revision of van der Waals atomic radii for molecular crystals: N, O, F, S, Cl, Se, Br, and I bonded to carbon. *Acta Crystallographica Section B: Structural Science*, *41*(4), 274–279.
33. Osetsky, Y. N., Bacon, D. J., Serra, A., Singh, B. N., & Golubov, S. I., (2003). One-dimensional atomic transport by clusters of self-interstitial atoms in iron and copper. *Philosophical Magazine*, *83*(1), 61–91.
34. Pauling, L. C., (1927). The theoretical prediction of the physical properties of many electron atoms and ions. Mole refraction, diamagnetic susceptibility, and extension in space. *Proceedings of the Royal Society of London. Series A, Containing Papers of a Mathematical and Physical Character*, *114*(767), 181–211.
35. Politzer, P., Murray, J. S., & Lane, P., (2003). Electrostatic potentials and covalent radii. *Journal of Computational Chemistry*, *24*(4), 505–511.
36. Rauscher, T., Dauphas, N., Dillmann, I., Fröhlich, C., Fülöp, Z., & Gyürky, G., (2013). Constraining the astrophysical origin of the p-nuclei through nuclear physics and meteoritic data. *Reports on Progress in Physics*, *76*(6), 066201.
37. Rusch, T., Lowry, P. B., Mair, P., & Treiblmaier, H., (2017). Breaking free from the limitations of classical test theory: Developing and measuring information systems scales using item response theory. *Information & Management*, *54*(2), 189–203.
38. Schomaker, V., & Stevenson, D. P., (1941). Some revisions of the covalent radii and the additivity rule for the lengths of partially ionic single covalent bonds. *Journal of the American Chemical Society*, *63*(1), 37–40.
39. Shannon, R. T., & Prewitt, C. T., (1970). Revised values of effective ionic radii. *Acta Crystallographica Section B: Structural Crystallography and Crystal Chemistry*, *26*(7), 1046–1048.
40. Sherby, O. D., & Taleff, E. M., (2002). Influence of grain size, solute atoms and second-phase particles on creep behavior of polycrystalline solids. *Materials Science and Engineering: A*, *322*(1–2), 89–99.
41. Smith, E., Evans, G., & Foadi, J., (2017). An effective introduction to structural crystallography using 1D Gaussian atoms. *European Journal of Physics*, *38*(6), 065501.

42. Su, C., Tripathi, M., Yan, Q. B., Wang, Z., Zhang, Z., Hofer, C., & Li, J., (2019). Engineering single-atom dynamics with electron irradiation. *Science Advances*, *5*(5), eaav2252.
43. Tansel, B., Sager, J., Rector, T., Garland, J., Strayer, R. F., Levine, L., & Bauer, J., (2006). Significance of hydrated radius and hydration shells on ionic permeability during nanofiltration in dead end and cross flow modes. *Separation and Purification Technology*, *51*(1), 40–47.
44. Walloch, P., Henke, B., Häuer, S., Bergmann, B., Spielmann, T., & Beitz, E., (2020). Introduction of scaffold nitrogen atoms renders inhibitors of the malarial L-lactate transporter, PfFNT, effective against the Gly107Ser resistance mutation. *Journal of Medicinal Chemistry*, *63*(17), 9731–9741.
45. Wang, L. W., & Zunger, A., (1994). Electronic structure pseudopotential calculations of large (approx. 1000 atoms) Si quantum dots. *The Journal of Physical Chemistry*, *98*(8), 2158–2165.

CHAPTER 7

The Electronic Structure of Atoms

Contents

7.1. INTRODUCTION

By adopting a conceptual framework wherein molecules are likened to elastic spheres, physicists have successfully anticipated and discussed certain observable phenomena at the macroscopic level, including the manifestation of gas pressure. Nevertheless, the model proposed by the researchers failed to incorporate the concept of molecular stability, specifically the underlying forces responsible for the cohesion of atoms (Thakkar, 2003). The recognition of the disparity between the governing laws of atoms and molecules, compared to those applicable to larger objects, was a protracted process marked by a prolonged acknowledgment period and subsequent acceptance. The commencement of a novel epoch in the field of physics can be traced back to the year 1900 when a youthful physicist of German origin named Max Planck emerged. The conventional belief among physicists was that energy is a continuous quantity, implying that radiation processes could release any arbitrary amount of energy. However, Planck's quantum theory challenged this longstanding assumption and brought about a significant paradigm shift in physics. Undoubtedly, the subsequent surge of research profoundly transformed the understanding of nature in a lasting manner. One crucial characteristic of a wave propagating through a medium is its velocity (u), which can be determined (Freed, 1971):

$$u = \lambda v$$

The multiplication of these factors yields units of distance per unit of time, commonly referred to as speed:

$$\frac{\text{distance}}{\text{time}} = \frac{\text{distance}}{\cancel{\text{wave}}} \times \frac{\cancel{\text{waves}}}{\text{time}}$$

The measurement of wavelength is commonly denoted in units of meters, millimeters, or nanometers, while frequency is typically quantified in hertz (Hz), where:

$$1\ Hz = 1\ cycle/s$$

The term "cycle" can be omitted, with the frequency denoted as, for instance, 25/s (pronounced as "25 per second") (Kim et al., 2018).

7.2. ELECTROMAGNETIC RADIATION

Many kinds of waves exist, including light, sound, and water. James Clerk Maxwell proposed that electromagnetic waves made up visible light in 1873.

According to Maxwell's theory, a magnetic and an electric field component are the two separate electromagnetic wave parts. The two components have similar speeds because they have the same wavelength and frequency. They do, however, spread out in planes that are perpendicular to one another. Maxwell's theory is significant because it offers a mathematical framework for understanding light's general properties and behavior. His model successfully explains how energy, in the form of radiation, can be transferred beyond the vastness of space using oscillating electric and magnetic fields (Kelly & Carter, 1980).

The process of creating and transmitting energy through electromagnetic wave propagation is called electromagnetic radiation. Electromagnetic waves move at about 3.00×10^8 meters per second in a vacuum, equal to 1,86,000 miles per second. Although the variations in velocity between different media are not large enough to significantly impact our computations, they exist. The velocity of electromagnetic waves, often known as the speed of light, is frequently denoted by the sign c in scientific domains. The wavelength of electromagnetic waves is a standard way to measure them, usually measured in nanometers (nm) (Sokolov et al., 2011).

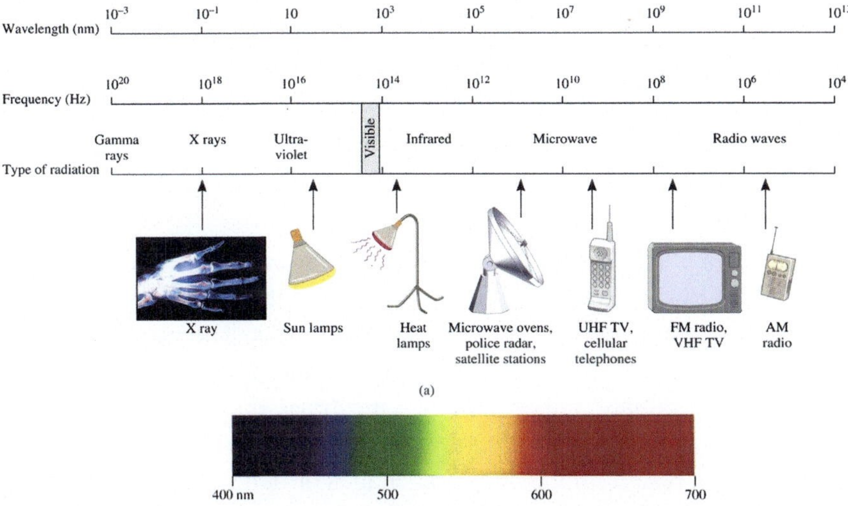

Figure 7.1. *(a) Classification of electromagnetic radiation. (b) The spectrum of electromagnetic radiation arranged in order of increasing wavelengths.*

Source: Martin, Creative Commons License.

The diagram presented in Figure 7.1 illustrates a range of electromagnetic radiation variants characterized by distinct variations in both wavelength and frequency. Large antennas, including those employed by broadcasting stations, emit long radio waves. The generation of shorter, visible light waves is attributed to the movements of electrons within molecules and atoms. Gamma rays, characterized by their short wavelength and high frequency, are generated due to nuclear transformations occurring within the atomic nucleus (Schneider et al., 2022).

Tip

Use visual aids, such as three-dimensional representations and electron density plots, to comprehend the shapes and orientations of atomic orbitals. This will enhance your understanding of electron distribution in atoms.

7.3. PLANCK'S QUANTUM THEORY

Measurements conducted during the latter portion of the 19th century revealed a correlation between the wavelength of an object and the quantity of radiant energy it emits at a given temperature. Explaining this interdependence using developed wave theory and thermodynamic principles yielded limited success. One theoretical framework discussed the correlation between short wavelengths but was inadequate in explaining the influence of longer wavelengths. Another theoretical framework successfully explained the phenomenon of longer wavelengths but proved inadequate in accounting for shorter wavelengths.

The laws of classical physics appeared to lack a fundamental component (Dirac, 1926). Planck addressed the problem by introducing an assumption that deviated significantly from established concepts. The classical physics framework operated under the assumption that atoms and molecules could emit or absorb radiant energy in any arbitrary quantity. According to Planck's postulation, atoms, and molecules can emit or absorb energy solely in discrete units, resembling compact packages or bundles. The terminology "quantum" was attributed by Planck to denote the minimum unit of energy that can be discharged (or assimilated) through electromagnetic radiation. The equation that describes the energy E of a single quantum of energy is expressed as follows (Bohr, 1932):

$$E = hv$$

The symbol "h" represents Planck's constant and is equal to 6.63 × 10^{-34} Joule-seconds, while "" denotes the frequency of radiation (Born, 2014).

$$E = \frac{hc}{\lambda}$$

As per the principles of quantum theory, energy is consistently emitted in discrete integer multiples of h. For instance, energy can be emitted as hv, 2 h, 3 h, and so forth, but not in non-integer multiples such as 1.67 hv or 4.98 hv. At the time of Planck's theory presentation, he encountered difficulties explaining the fixed or quantized nature of energies.

Commencing from this hypothesis, the researcher encountered no difficulty establishing a correlation between the experimental data about the emission by solids across the complete spectrum of wavelengths. Notably, all the collected data consistently substantiated the principles of the quantum theory (Herschbach, 2001). The notion that energy ought to be quantized or discretized may appear peculiar, yet the principle of quantization exhibits numerous analogies.

As an illustration, it is worth noting that the quantization of electric charge is observed, wherein solely integral multiples of e, representing the charge of a single electron, are permissible. The quantization of matter arises from the requirement that the quantities of electrons, protons, neutrons, and atoms within a given sample of matter must be whole numbers. Quantized phenomena are observed in various processes within living systems. The eggs produced by hens exhibit discrete characteristics, while a pregnant feline gives birth to whole numbers of kittens rather than fractional or fractional multiples of a kitten (Agullo et al., 2013).

Remember

According to quantum mechanics, electrons can only exist in discrete energy levels. When an electron changes energy levels, it does so instantly, skipping over the intervening levels entirely. One common term for this event is "quantum leap."

7.4. THE PHOTOELECTRIC EFFECT

In the year 1905, a mere half-decade after the introduction of Max Planck's quantum theory, the esteemed physicist Albert Einstein, of German–American descent, employed said theory to successfully discuss an additional phenomenon within the field of physics known as the photoelectric effect.

This phenomenon entails the expulsion of electrons from the surface of specific metals upon exposure to light possessing a frequency surpassing a designated threshold frequency (Pratt, 1960). The number of electrons emitted exhibited a direct relationship with the intensity, or luminosity, of the light, while the energies of the emitted electrons did not display such proportionality. No electrons were emitted regardless of the light's intensity when the frequency was below the threshold (Eichler & Stöhlker, 2007).

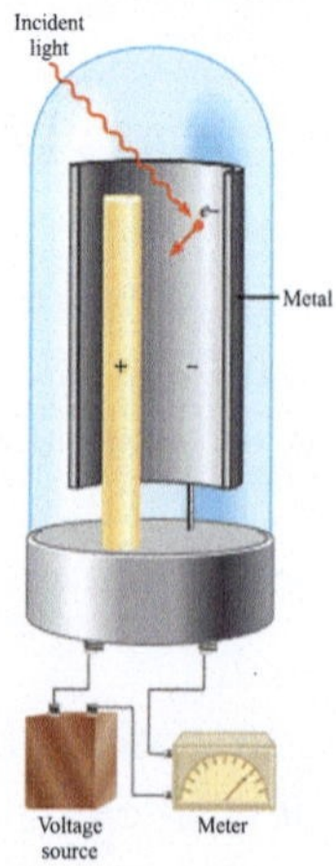

Figure 7.2. *An apparatus for studying the photoelectric effect.*

Source: Linus, Creative Commons License.

The wave theory of light was insufficient in explaining the observed phenomena of the photoelectric effect. Einstein, however, posited an exceptional assumption. He proposed that a beam of light can be conceptualized as a series of discrete particles. The term "photons" is currently used to refer to these individual light units (Figure 7.2). Einstein derived (Ossiander et al., 2018):

$$E = hv$$

Did You Know?

Contrary to popular depictions of electrons orbiting the nucleus like planets around the sun, electrons in atoms are best described as existing in a probability cloud. These electron clouds represent regions where electrons are most likely to be found, based on their wave-like behavior.

7.5. BOHR'S THEORY OF THE HYDROGEN ATOM

The contributions made by Einstein laid the foundation for resolving an additional enigma in the field of physics during the 19th century, specifically about the emission spectra exhibited by atoms (Suto, 2021) (Figure 7.3).

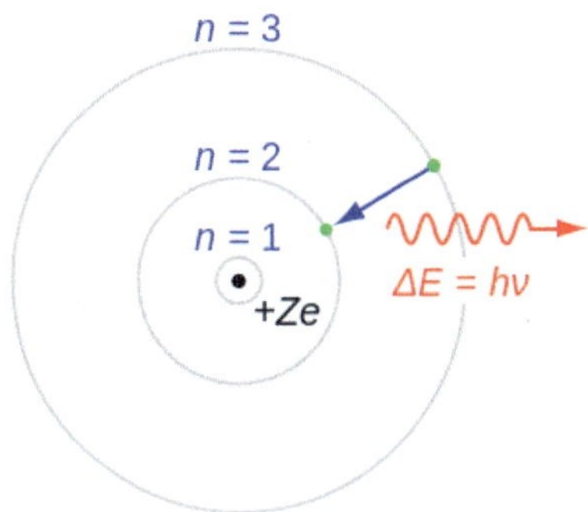

Figure 7.3. *Bohr's cake model of hydrogen atom.*

Source: Wikipedia, Creative Commons License.

7.5.1. Emission Spectra

Since the 17th century, scholars have been actively involved in the investigation of emission spectra, which pertain to the radiation produced by substances in continuous or line spectra. The inception of this area of study can be attributed to Newton's experimental evidence, which established that sunlight is composed of distinct color constituents that can be amalgamated to produce white light. After their discovery, chemists, and physicists explored and analyzed the properties and attributes exhibited by these emission spectra. Observing an object's emission spectrum involves exposing a material sample to thermal energy or an alternative energy source. In the case of a gaseous substance, this can include subjecting it to a high-voltage electrical discharge. (Singh, 2012).

An iron bar recently extracted from a high-temperature source exhibits a discernible luminosity, commonly known as "red-hot" or "white-hot." The observable brightness is synonymous with the portion of its electromagnetic spectrum perceived by the human visual system. The thermal energy emitted by the iron bar corresponds to an additional segment of its electromagnetic spectrum, specifically falling within the infrared region. Both the emission spectra of the sun and a heated solid display a common attribute known as continuity, whereby the spectra encompass all wavelengths of visible light (Bopegedera, 2011).

In contrast, the emission spectra of atoms in the gaseous phase do not display a continuous range of wavelengths extending from red to violet. In contrast, atoms produce discrete and distinguishable bright lines in different segments of the visible spectrum.

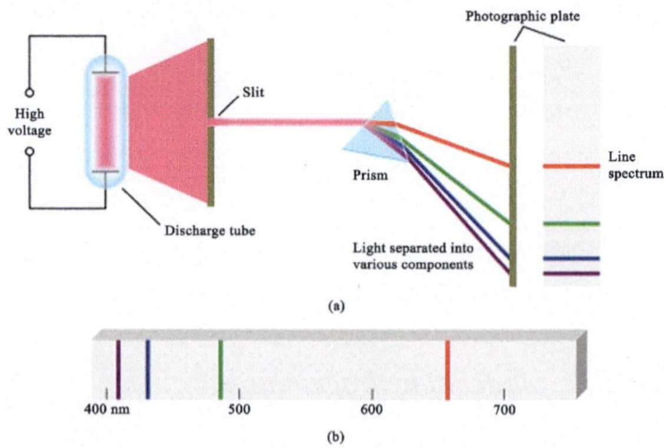

Figure 7.4. *(a) Experimental setup for studying the emission spectra of atoms and molecules. (b) The spectral lines observed in emission spectra.*

Source: Jack, Creative Commons License.

The line spectra that have been observed indicate the emission of light occurring solely at specific and separate wavelengths. Every individual element exhibits a unique and distinctive emission spectrum (Shields & Kash, 1992). Employing characteristic lines observed in atomic spectra carries substantial importance in chemical analysis, as it facilitates the identification of unfamiliar atoms, akin to using fingerprints for identifying individuals. The confirmation of the identification of an unknown sample occurs when the emission spectrum lines of the sample precisely match the emission spectrum lines of a known element. This procedure's effectiveness was recognized in chemical analysis, although its origin remained unknown until the early 20th century (Clark & Chamberlain, 2014) (Figure 7.4).

7.5.2. Emission Spectrum of the Hydrogen Atom

In 1913, shortly following the groundbreaking discoveries made by Planck and Einstein, the Danish physicist Niels Bohr put forth a theoretical explanation of the emission spectrum exhibited by the hydrogen atom. A high level of complexity characterizes Bohr's theoretical framework and is presently regarded as incomplete. Therefore, considering the spectral lines, the focus will be solely on the author's significant assumptions and ultimate findings (Zhou & Yuan, 2008).

Hence, it can be inferred that an electron of this nature would undergo rapid inward spiraling towards the nucleus, ultimately resulting in self-annihilation upon interaction with the proton. To explain the absence of this phenomenon, Bohr proposed the concept that the electron is restricted to occupying discrete orbits characterized by distinct energy levels. Put differently, the energy levels of electrons exhibit quantization. An electron within any permissible orbit will not emit energy, thereby preventing its gradual descent toward the atomic nucleus (Fujita et al., 1971). Bohr showed that the energy levels available to an electron within a hydrogen atom are determined by:

$$E_n = -R_H\left(\frac{1}{n^2}\right)$$

The Rydberg constant, denoted as RH and attributed to the Swedish physicist Johannes Rydberg, is quantified as 2.18×10^{-18} J for the hydrogen atom. The primary quantum number, denoted as n, is a discrete integer ranging from 1 to infinity. The conventional method involves incorporating a negative sign in the equation above to indicate that the energy of an electron within an atom is considerably lower than that of a free electron situated at an infinite distance from the atomic nucleus (Mills, 2000). In conventional terminology, a free electron is commonly described as possessing zero energy. Mathematically speaking, the process described can be interpreted as substituting an infinite value for n in the equation mentioned earlier, leading to the conclusion that E becomes zero. The magnitude of the electron's energy, denoted as En, increases while also exhibiting a greater deviation towards the negative side as the distance between the electron and the nucleus diminishes, corresponding to a decrease in the principal quantum number, n (Singleton et al., 2022).

At n = 1, the system attains its minimum energy state, associated with the highest negativity and stability. Within this chapter, the terminology "ground state" or "ground level" is employed to denote the state of minimal energy within a specified system, particularly in the context of an atom. The stability of the electron is diminished as the primary quantum number, denoted as n, increases beyond a value of 1. Each level is referred to as an excited state or excited level due to higher energy than the ground state. An electron in a hydrogen atom is considered excited when its principal quantum number (n) exceeds 1. Determining the radius for each circular orbit in Bohr's model relies on the value of n^2. Hence, the orbit radius undergoes a substantial and rapid expansion as the value of n increases successively from 1 to 2 to 3. The strength of the electron-nucleus binding diminishes as the electron transitions to higher energy states due to the concurrent increase in their separation distance (Barbara et al., 1989).

Bohr's theory offers a fundamental framework for explaining the phenomenon of line spectrum observed in the hydrogen atom. The phenomenon in which an electron moves from a lower energy state characterized by a smaller primary quantum number (n) to a higher energy state characterized by a larger value of n is attributed to the absorption of radiant energy by the atom. Conversely, when an electron transitions from an excited state to a lower energy state, it emits radiant energy as a photon. The phenomenon of an electron transitioning between discrete energy levels can be aptly compared to a tennis ball's movement along a flight of steps. The ball can assume various discrete positions or steps but cannot occupy an intermediate position among these steps. Transitioning from a lower energy state to a higher energy state necessitates the input of energy, whereas the reverse transition from a higher energy state to a lower energy state results in the release of energy (Yuan et al., 2023). The spatial distance between the initial and final stages determines the energy required for each type of transformation. The quantum mechanical transition of an electron within a Bohr atom is contingent upon the disparity in energy levels between its initial and final states. This difference serves as the determining factor for the quantum of energy required for such electronic movement.

Applying the equation above to the emission process occurring in a hydrogen atom, it is postulated that the electron initially occupies an excited state, represented by the principal quantum number n*i*. The primary quantum number n*f* indicates the transition of an electron from a higher energy state to a lower energy state during the emission process. The subscripts i and f represent the initial and final states, respectively. Two potential manifestations of the lower energy state are the fundamental ground state and a reduced excitation state.

$$\Delta E = E_f - E_i$$

From the previous equation,

$$E_f = -R_H\left(\frac{1}{n_f^2}\right)$$

And

$$E_i = -R_H\left(\frac{1}{n_i^2}\right)$$

Therefore,

$$\Delta E = \left(-\frac{R_H}{n_f^2}\right) - \left(-\frac{R_H}{n_i^2}\right) = R_H\left(\frac{1}{n_i^2} - \frac{1}{n_f^2}\right)$$

or,

$$\Delta E = hv = R_H\left(\frac{1}{n_i^2} - \frac{1}{n_f^2}\right)$$

The energy level (n*i*) at the beginning of the photon emission process is higher than the energy level (n*f*) at the end. Consequently, the enclosed word within parentheses possesses a negative value, indicating that E is likewise negative and signifies a depletion of energy in the surrounding environment. The enclosed term indicates a positive value, indicating that the energy change (E) is positive when energy is absorbed, and the initial energy level (n*i*) is lower than the final energy level (n*f*). Each unique spectral line observed in the emission spectrum can be attributed to a specific transition within a hydrogen atom. The investigation of a significant number of hydrogen atoms allows for the examination of the complete spectrum of transitions that occur, resulting in the formation of corresponding spectral lines. The intensity of a spectral line is determined by the number of photons emitted at a specific wavelength (Verner et al., 1999).

Each sequential transition recorded in the hydrogen spectrum is included in Table 7.1, along with the name of the individual who discovered it. Because several of the Balmer series' spectral lines are located inside the visible spectrum, it has been determined that this series is especially suitable for examination.

Table 7.1. The Various Series in the Atomic Hydrogen Emission Spectrum

Series	**n_f**	**n_i**	**Spectrum Region**
Lyman series	1	2,3,4, and so on	UV
Balmer series	2	3,4,5, and so on	Visible and UV

Paschen series	3	4,5,6, and so on	IR
Brackett series	4	5,6,7, and so on	IR

Source: Phillip, Creative Commons License.

However, it is more useful to describe transitions as demonstrated below. Each horizontal line in the diagram represents a discrete energy level permissible for the electron within a hydrogen atom (Carrington, 1964) (Figure 7.5).

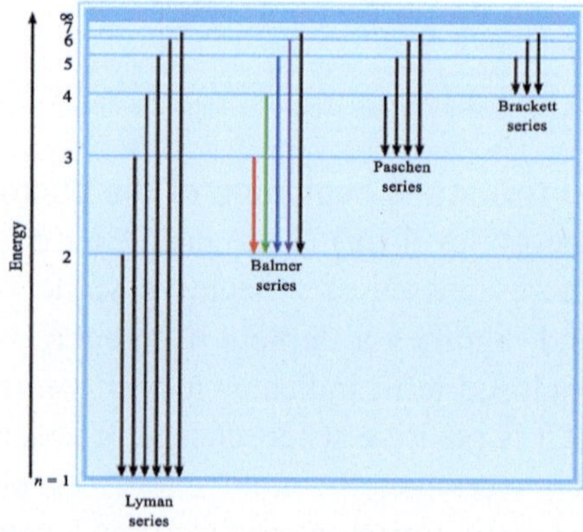

Figure 7.5. *The energy levels in the hydrogen atom and the various emission series.*

Source: Philip, Creative Commons License.

Did You Know?

Quantum mechanics allows for a remarkable phenomenon called electron tunneling. When a barrier separates two regions, like in certain electronic devices, electrons can "tunnel" through the barrier without having enough energy to surmount it. This plays a crucial role in phenomena like scanning tunneling microscopy.

7.6. THE DUAL NATURE OF THE ELECTRON

Bohr's theory elicited both perplexity and fascination among physicists. The inquiry was raised regarding quantizing energy levels in the hydrogen electron.

Alternatively, rephrasing the inquiry more precisely, what is the underlying reason for the confinement of the electron within specific, predetermined orbits around the nucleus in a Bohr atom? For 10 years, there was a lack of a coherent explanation, with even Bohr unable to provide one. In 1924, Louis de Broglie, a physicist from France, presented a resolution to this enigma. De Broglie postulated that if light waves can exhibit particle-like behavior in the form of photons, it is conceivable that particles like electrons could also exhibit wave-like characteristics (Figure 7.6).

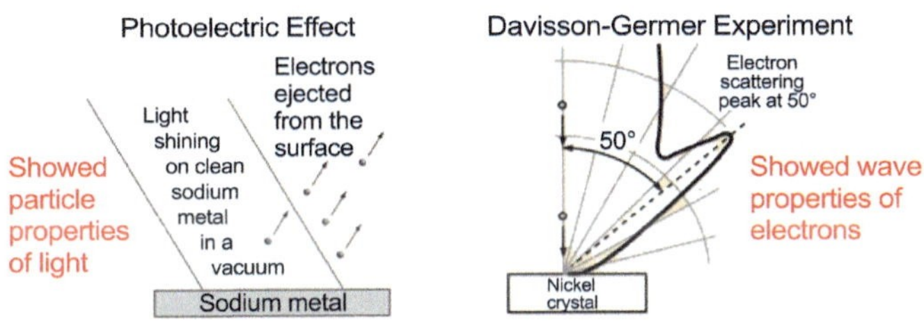

Figure 7.6. *Illustration of wave-particle (dual) nature of electrons.*

Source: Hyper Physics, Creative Commons License.

As per de Broglie's postulation, an electron confined within the nucleus exhibits characteristics akin to a standing wave. The act of plucking a guitar string has the potential to generate standing waves (Gretarsson et al., 2011). The waves are characterized as standing or stationary due to their resemblance of propagation along a string.

Certain nodes on the string remain stationary, exhibiting zero amplitude, indicating that the wave does not propagate at these specific locations. Each terminal is associated with a node; additional nodes can be located between the terminals. A direct relationship exists between the frequency of vibration and the wavelength of the standing wave, whereby an increase in frequency results in a decrease in wavelength. Additionally, a higher vibration frequency corresponds to more nodes in the standing wave (Picozzi et al., 2007).

De Broglie postulated that in a hydrogen atom, if an electron's behavior seems like that of a standing wave, the wavelength of the wave must precisely correspond to the circumference of the electron's orbit. Alternatively, the wave would undergo partial cancellation during each subsequent orbit. Ultimately, the magnitude of the wave would diminish to null, resulting in the cessation of the wave's existence (Stamplecoskie et al., 2015) (Figure 7.7).

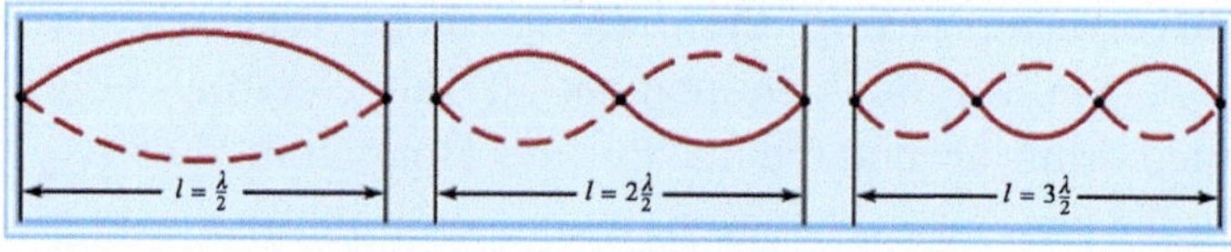

Figure 7.7. *The standing waves are produced by plucking any guitar string.*

Source: Phillip, Creative Commons License.

Mathematically:

$$2\pi r = n\lambda$$

where; r is the radius of the orbit; "λ" is the wavelength of the electron wave; and "n" is an integer starting from 1 and increasing. The quantization of the electron's energy results from its dependence on the orbit size, specifically the value of r. De Broglie's line of reasoning yielded the inference that waves possess the capacity to manifest particle-like characteristics, while particles can demonstrate properties akin to those of waves. De Broglie's deduction posited a correlation between the properties of particles and waves, as expressed by an equation (Celano et al., 2015) (Figure 7.8):

$$\lambda = \frac{h}{mu}$$

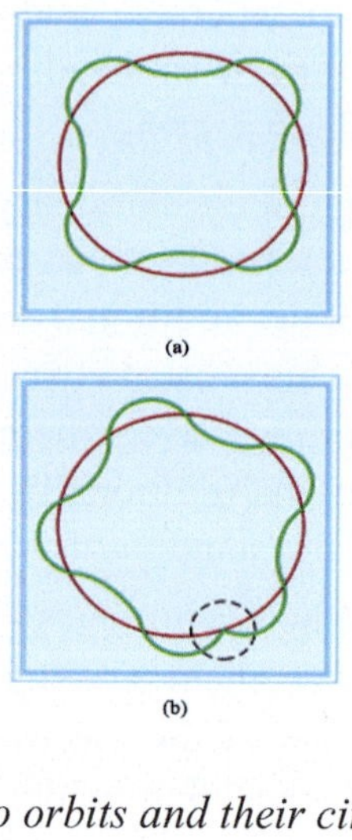

Figure 7.8. *Illustration of two orbits and their circumferences.*

Source: Gary, Creative Commons License.

7.7. QUANTUM MECHANICS

After the amazing success of Bohr's hypothesis, a string of disappointments followed. Bohr's method proved insufficient to explain the emission spectra seen in atoms containing multiple electrons, such as helium and lithium atoms. A wave's spatial expansion makes it difficult to pinpoint its exact location. Electrons' dualistic properties presented considerable difficulties due to their incredibly low mass.

Werner Heisenberg, a well-known physicist from Germany, tackled the problem of establishing the position of a subatomic particle that shows wave-like behavior. He developed a theory known as the Heisenberg uncertainty principle. According to this principle, properly determining a particle's momentum and location is impossible (Marcus, 1966).

Erwin Schrödinger, an Austrian scientist, created an equation in 1926 using a challenging mathematical method. This equation can be compared to Newton's laws of motion for macroscopic objects and is used to generalize the behaviors and energy of submicroscopic particles. Advanced mathematics is required to solve the Schrödinger equation, although this discussion will not cover that subject in detail.

It is important to understand that the wave function, which depends on the spatial location of the system (for example, an electron within an atom), and the mass variable (m), which represents particle properties, are both included in the equation (Neese et al., 2019).

The wave function lacks inherent physical significance. Nevertheless, the magnitude of the wave function squared ψ^2 directly correlates with the probability of locating the electron within a specific spatial domain. The concept of relating the square of the wave function to probability was initially developed based on an analogy derived from wave theory. Based on wave theory, one can posit that the square of a wave's amplitude directly correlates with the magnitude of the wave.

The likelihood of locating an electron near the atomic nucleus is connected to the square of the wave function. The introduction of Schrödinger's equation in physics and chemistry marked a significant milestone as it paved the way for the emergence of a novel domain known as quantum mechanics, or wave mechanics. The term "old quantum theory" is commonly employed in contemporary research to denote the temporal span from 1913, when Bohr initially introduced his investigation of the hydrogen atom, to 1926 (Bader, 2005).

7.7.1. The Quantum Mechanical Description of the Hydrogen Atom

The Schrödinger equation, a fundamental equation in quantum mechanics, describes the energy levels at which an electron can exist within a hydrogen atom. Furthermore, it provides the corresponding wave functions that delineate the probability distribution of detecting the electron in different physical locations. The subsequent discussion will offer a comprehensive analysis of a specific set of quantum numbers that can be employed to depict the hydrogen atom's energy levels and wave functions (Sakimoto, 2001). The comprehensive model of the hydrogen atom can be constructed due to the presence of quantum numbers. As per the principles of quantum mechanics, the precise position of an electron within an atom is inherently uncertain. However, it delineates the spatial region within which the electron is most probable to be located during a specific temporal interval. The concept of electron density provides insight into the likelihood of locating an electron within a specific region of an atom. The square of the wave function denoted as Ψ^2, determines the spatial distribution of electron density in a three-dimensional region surrounding the atomic nucleus. A region with a higher electron density indicates a greater likelihood of electron presence, whereas a region with a lower electron density indicates a decreased tendency for electron presence (Casati et al., 1987) (Figure 7.9).

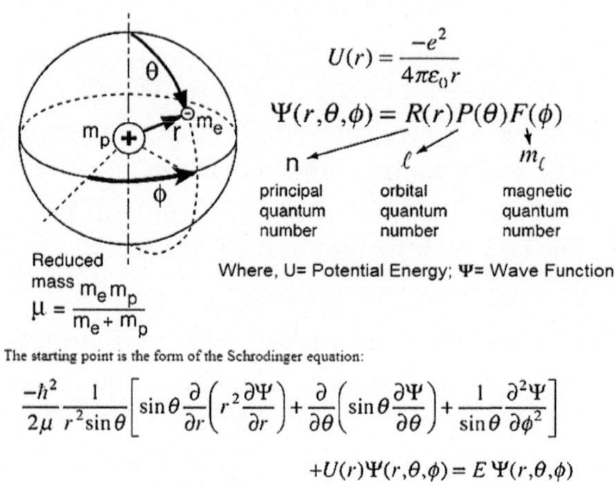

Figure 7.9. *Illustration of Schrodinger equation for hydrogen atom.*

Source: Hyperphysics, Creative Commons License.

The terminology employed pertains to an atomic orbital instead of an orbit, thereby differentiating the quantum mechanical depiction of an atom from Bohr's theoretical construct. An atomic orbital can be conceptualized as the mathematical representation of the wave function associated with an electron confined within an atom. The term "orbital" denotes a specific region in which an electron is located. The localization of an electron in a specific area is determined by the square of the wave function associated with the relevant orbital, which provides information on the electron density or the probability of locating the electron. Consequently, every atomic orbital possesses a distinct energy level and electron density distribution (Gaeta & Stroud, 1990). The Schrödinger equation demonstrates satisfactory performance when employed to characterize the behavior of the hydrogen atom, consisting of a solitary proton and a solitary electron. Nevertheless, it has been discovered that the precise solution to the Schrödinger equation is unattainable for atoms containing multiple electrons. Fortunately, chemists and physicists have developed strategies to overcome this challenge using approximation methodologies. The distinction between the behavior of electrons in many-electron atoms and hydrogen atoms is widely recognized. Nevertheless, it is widely held that the disparity between the two will likely be minimal. To gain a deeper understanding of the behavior of electrons in intricate atomic systems, it is possible to utilize the energies and wave functions derived from the hydrogen atom as reliable approximations. Indeed, this methodology offers reasonably dependable depictions of the electrical characteristics of atoms containing multiple electrons (Hinton et al., 1992).

7.8. QUANTUM NUMBERS

In accordance with the principles of quantum mechanics, an accurate depiction of electron distribution in hydrogen and other atoms necessitates the utilization of three quantum numbers. The mathematical derivation of the Schrödinger equation for a hydrogen atom resulted in the aforementioned numerical values. The primary quantum number, the angular momentum quantum number, and the magnetic quantum number are employed in the solution of the equation. Quantum numbers characterize atomic orbitals and assign identities to the electrons occupying them. A fourth quantum number, known as the spin quantum number, discusses the characteristics and actions of individual electrons, leading to a comprehensive depiction of electrons within atomic systems (Harth et al., 1989).

7.8.1. The Principal Quantum Number (n)

Assigning integral values like 1, 2, 3, and so on to the main quantum number n is common practice. The energy contained within an orbital of a hydrogen atom is controlled by the fundamental quantum number, which is written as n. In the debate that is to follow, it will be demonstrated that this assertion is untrue for an atom that possesses a greater number of electrons. Within a particular orbital, the main quantum number is also linked to the mean radial distance that separates the electron from the nucleus. When the value of n increases, the average distance that an orbital electron is from the nucleus also increases, increasing the orbital size (Kutzelnigg, 2008).

7.8.2. The Angular Momentum Quantum Number (

The angular momentum quantum number, denoted as l, is responsible for illuminating the geometric characteristics of the orbitals. The values of the primary quantum number, n, determine the values of the other variables. When n is given a certain value, the integral range of the variable l is from 0 to n–1 (Krems et al., 2004). When n equals 1, there is only one possible value for l, which can be discovered by subtracting 1 from n, which gives l the value of 0, the only valid option. When the value of n is set to 2, the variable l can take on two distinct forms: 0 or 1. When the value of n is set to 3, three distinct values of l can be represented by the numbers 0 through 2. These values are 0, 1, and 2. The value of l is usually denoted by the letters s, p, d, etc. Other letters, such as c, m, and n, may also be used (Kraft et al., 1990):

ℓ	0	1	2	3	4	5
Name of orbital	*s*	*p*	*d*	*f*	*g*	*h*

Therefore, the formation of a s orbital occurs when the value of l is equal to zero. Like the previous example, a p orbital is generated when the value of l is set to 1, and so on. A collection of orbitals with the same primary quantum number is called a shell. The letter n denotes this number. A subshell is a collection of one or more orbitals that share the same primary quantum number (n) and azimuthal quantum number (l) as one another (Aquilanti & Grossi, 1980). This concept can be illustrated in the shell with a fundamental quantum number (n) of two and containing two subshells with l = 0 and l = 1, respectively. These values are the variations for n = 2 that can be considered valid alternatives. The in-question subshells have been given the names of the 2s and 2p subshells, respectively. The number 2 stands in for the primary quantum number, which is written as n, while the letters s and p stand in for the azimuthal quantum numbers, which are written as l (Arbó et al., 2008).

Remember

Electrons can become quantum-entangled with each other, which means that their quantum states become correlated and interdependent, even when separated by vast distances. This concept is at the heart of quantum computing and quantum information theory.

7.8.3. The Magnetic Quantum Number

The magnetic quantum number, denoted by the letter m, is the property that identifies and describes the spatial orientation of an orbital. The value of the magnetic quantum number within a subshell is determined by the value of the angular momentum quantum number, denoted by the letter l. As seen in the following table, there are two and a half plus one integral values of m for any given value of l (Mulliken, 1932).

If the quantum number l is equal to zero, the magnetic quantum number m_l likewise equals 0 because this is the only possible outcome. It is possible to obtain any one of the three possible values of m_l that are present if the value of l is equal to 1, specifically "–1," "0," and "1." The number of possible values for m_l can be determined by evaluating the phrase "2 × 2 + 1" when l is equal to 2. This will give the total number of possible combinations for m_l. As a result, the following five values are obtained: –2, –1, 0, 1, and 2. The ratio of the number of orbitals to the number of m l values in a subshell is directly proportional to the value of the l parameter (Pauling, 1927).

In conclusion, it is important to consider the case in which the main quantum number (n) equals 2, but the azimuthal quantum number (l) equals 1. The fact that the parameters n and l have been assigned gives the impression that the atomic structure includes a 2p subshell. Because three distinct 2p orbitals are included within this subshell, the variable m_l can take on one of three possible values: –1, 0, or 1 (Kolbinger et al., 2021).

7.8.4. The Electron Spin Quantum Number (m_s)

Experiments were performed on the emission spectra of hydrogen and sodium atoms, and the results showed that the lines within these spectra may be separated into several components when an external magnetic field is applied. This was discovered by observing the effects of the experiments on the spectra. Physicists were forced to postulate that electrons display characteristics comparable to those of microscopic magnets to explain these results (Gálvez et al., 2019). It is possible to explain the magnetic properties of electrons by considering their hypothetical rotation on individual axes, which is analogous to the rotation of the Earth. The rotating motion of a charged particle, as described by the principles of electromagnetic theory, generates a magnetic field, which, in turn, leads to the development of magnetic properties in an electron. Figure 7.10 demonstrates the two distinct rotational movements that an electron is capable of exhibiting. These movements are known as clockwise and counterclockwise rotations (Auböck et al., 2008).

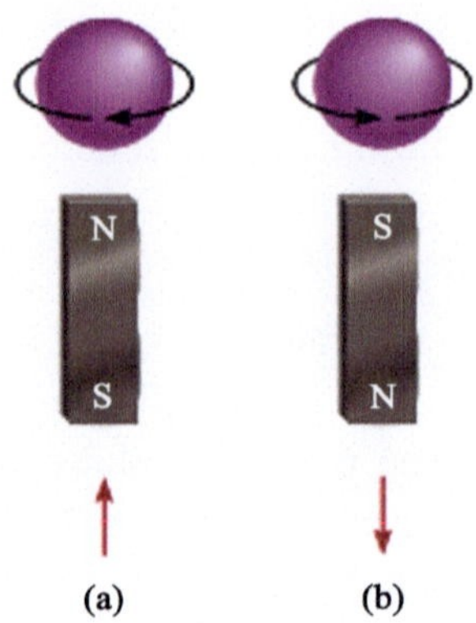

Figure 7.10. *The (a) clockwise; and (b) counterclockwise spins of an electron.*

Source: Warren, Creative Commons License.

7.9. ATOMIC ORBITALS

It has been noticed that the expression (2*l* + 1) evaluates to 1 when the quantum number *l* is equal to 0, which results in the existence of just one permissible value for the magnetic quantum number ml. As a direct result, the configuration corresponds to a s orbital. The value 3 is obtained by evaluating the formula (2*l* + 1) when the quantum number *l* equals 1. Consequently, there are three distinct values of m, corresponding to three different p orbitals represented by the symbols px, py, and pz, respectively. Because the expression (2*l* + 1) evaluates to 5 when the quantum number *l* is equal to 2, it may be deduced that there are five distinct possibilities for the value of ml. These five values of ml are connected to the d orbitals, which are indicated by more specific subscripts (Tsaparlis, 1997).

7.9.1. s-Orbitals

The analysis of atomic orbitals raises several fundamental questions, the most important of which concerns determining the orbitals' forms. Because the wave function associated with an orbital spread from the nucleus to an infinite distance, one can say that an orbital does not have a clearly defined shape from a strict point of view. This is because the wave function extends in all directions from the nucleus. Determining what an orbital's visual depiction should look like is difficult. However, conceiving orbitals as bearing unique geometries when forming chemical bonds between atoms is unquestionably beneficial (Raffenetti, 1973).

Empirical evidence reveals that electrons typically occupy regions close to the nucleus. Even though it is theoretically conceivable for an electron to exist in any location, it is more common for electrons to inhabit locations close to the nucleus. The following image illustrates the spatial distribution of electron density within a hydrogen 1s orbital as the orbital moves away from the atomic nucleus (Knizia, 2013). As one moves further and further out from the nucleus, it is plain to see that the electron density drops substantially. It is estimated that there is a likelihood of roughly 90% that the electron is situated within a sphere that surrounds the nucleus and has a radius of 100 picometers (1 picometer is equal to 1×10^{-12} meters). Therefore, a boundary surface diagram that includes about 90% of the total electron density within the 1s orbital can be used to represent the 1s orbital visually. This diagram can be used to portray the 1s orbital visually.

A simple spherical shape can be used to depict a 1s orbital when represented in this manner (Richardson et al., 1962). The s orbitals have a spherical shape, but their sizes can vary, and it has been shown that these sizes grow larger when the primary quantum number increases. This constraint does not present a significant disadvantage, although it is impossible to identify the specific electron density fluctuations within each barrier surface. The fundamental aspects of atomic orbitals that are significant to humans are their geometric configurations and relative magnitudes. Boundary surface diagrams can efficiently describe these aspects of atomic orbitals (Nakiboglu, 2003) (Figure 7.11).

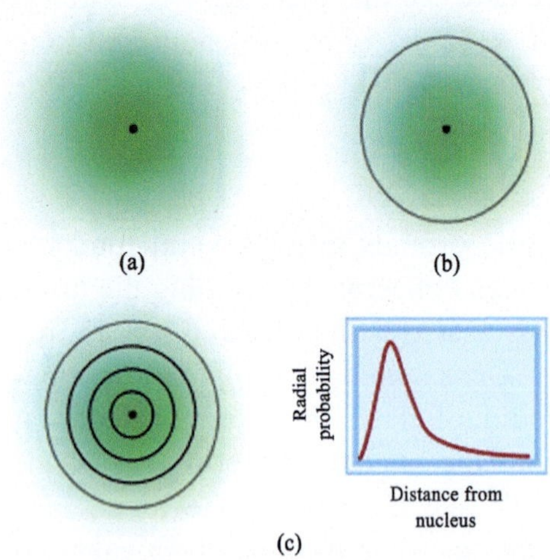

Figure 7.11. *(a) A plot showing the electron density in the hydrogen 1s orbital relative to the nucleus's distance. (b) A schematic of the hydrogen 1s orbital's boundary surface. (c) Dividing the 1s orbital into successive spherical thin shells will allow you to see the electron density distribution more realistically.*

Source: Warren, Creative Commons License.

7.9.2. p-Orbitals

Evidently, at the fundamental quantum number n = 2, the p orbitals start. The only possible value for the angular momentum quantum number when n is one is zero, resulting in a single 1s orbital formation. As stated before, when the orbital quantum number l is 1, the magnetic quantum number m may take on the values of –1, 0, and 1. Starting with the initial values of n = 2 and l = 1, there are three 2p orbitals total—2px, 2py, and 2pz. The letter subscripts denote the axis that the orbitals are directed across. The three p orbitals are identical in size, shape, and energy, except for their differing inclinations. It should be noted that there is no direct correlation between the values of the variable m and the axes of x, y, and z. Because there are three possible values of ml, it is imperative to remember that there are three p orbitals with distinct orientations (Zhao et al., 2012) (Figures 7.12 and 7.13).

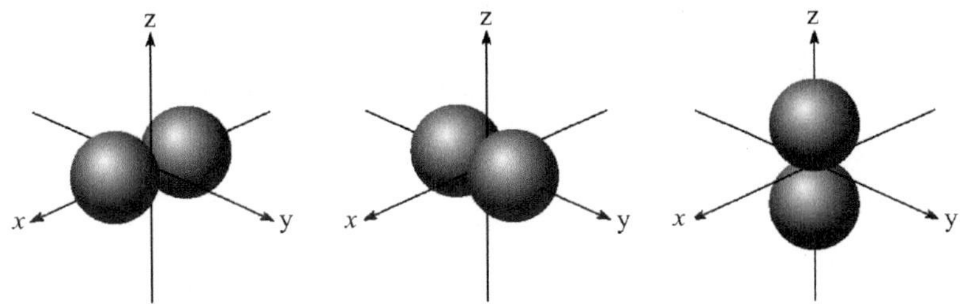

Figure 7.12. *The boundary surface diagrams of the three 2p orbitals. These orbitals are identical in shape and energy but differ in orientations.*

Source: Warren, Creative Commons License.

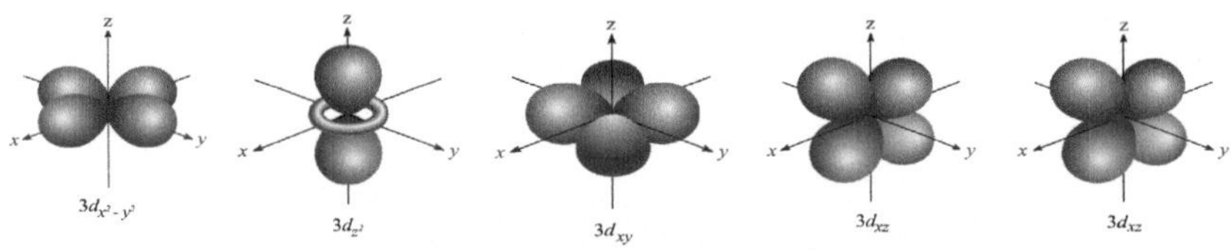

Figure 7.13. *Boundary surface diagrams of the five 3d orbitals.*

Source: Roald, Creative Commons License.

7.9.3. d-Orbitals and Other Higher-Energy Orbitals

Five different values of ml correspond to the five d orbitals when the value of l equals 2. For a d orbital, n must be at least 3. It is established that l cannot be greater than n–1 because of the fundamental nature of the relationship between l and n (Barnum, 1961). There are five 3d orbitals in the particular case where n = 3 and l = 2, namely 3dxy, 3dyz, 3dxz, $3dx^2$ y^2, and 3dz. The d orbitals display a variety of orientations that correspond to various ml values, much like the p orbitals do (Aikens, 2018).

Nevertheless, it's vital to remember that there is no certain relationship between an orientation and a m value. All of an atom's third-order orbitals have the same energy levels. Similar geometries can be seen in the d orbitals with 4d, 5d, and higher principal quantum numbers (Shen et al., 2022).

The alphabetical labels f, g, and subsequent refer to orbitals with energy levels above d orbitals. The f orbitals are crucial in explaining the properties of elements with atomic numbers over 57, but illustrating them can be difficult due to their intricate forms. In general chemistry, attention does not extend to orbitals with quantum numbers higher than 3, notably the g orbitals and beyond (Howell et al., 1973).

7.9.4. The Energies of Orbitals

Having acquired a basic comprehension of the geometric characteristics and dimensions of atomic orbitals, it is time to investigate their comparative energy levels and examine the influence of energy levels on the specific configuration of electrons within atoms (Senet, 1997).

The energy of an electron in a hydrogen atom is exclusively determined by its principal quantum number, as indicated by the equation provided. Consequently, the energy levels of hydrogen orbitals exhibit an ascending order as follows (Ayers et al., 2005):

$$1s < 2s = 2p < 3s = 3s = 3p = 3d < 4s = 4p = 4d = 4f < \cdots$$

It is crucial to note that the electron's energy in hydrogen remains constant regardless of whether it occupies the 2s or 2p orbital, despite the discernible changes in electron density distributions between the 2s and 2p orbitals. A hydrogen atom's 1s orbital is related to the ground state, which denotes the configuration with the greatest stability. Due to its proximity to the atomic nucleus, the electron in this orbital has the strongest electrostatic attraction with the nucleus. An electron in a hydrogen atom's 2s, 2p, or any higher orbital is thought to be excited (Shirley, 1972).

Compared to hydrogen, the complexity of the energy landscape is much higher for atoms containing multiple electrons. An electron's primary and angular momentum quantum numbers in an atom define its energy. According to empirical observations, the energy level corresponding to the 3d orbital and the energy level corresponding to the 4s orbital are physically close in atoms with multiple electrons. An atom's total energy is determined by the sum of all of its orbitals' energies and by the energy produced by the repulsion between the electrons occupying these orbitals. According to research, when the 4s subshell is filled before the 3d subshell, the total energy of an atom is reduced (Ozaki, 2003) (Figures 7.14 and 7.15).

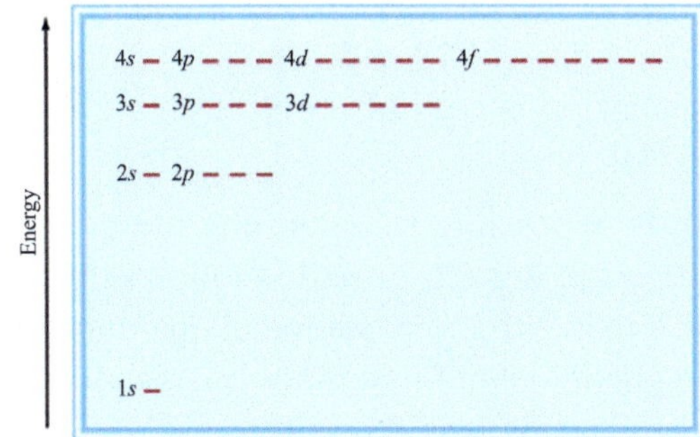

Figure 7.14. *Levels of orbital energy in the hydrogen atom. Each thin horizontal line shows one orbital. The energy of all orbitals with the same primary quantum number (n) is the same.*

Source: Roald, Creative Commons License.

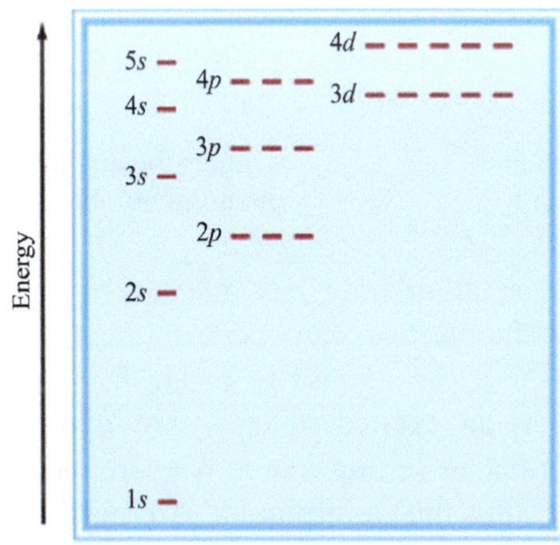

Figure 7.15. *Levels of orbital energy in an atom with numerous electrons. Be aware that the values of n and l affect the energy level.*

Source: Roald, Creative Commons License.

7.10. ELECTRON CONFIGURATION

Using the four quantum numbers *n, l, ml*, and ms facilitates the thorough labeling of an electron within any orbital present in an atom. Similar to how a person's location is characterized by a combination of their street address, city, state, and postal ZIP code, the set of four quantum numbers can describe the spatial properties of an electron within an atom (Vander Sluis & Nugent, 1974).

For illustration, it is possible to determine the quantum numbers that apply to an electron occupying a 2s orbital as follows: the principal quantum number (n) is given a value of 2, the azimuthal quantum number (*l*) is given a value of 0, the magnetic quantum number (ml) is given a value of 0, and the spin quantum number (ms) can take either +1/2 or –1/2.

The trouble of painstakingly writing each quantum number is the motivation for using simplified notation. The quantum numbers in the example above can be written as (2, 0, 0, +1/2) or (2, 0, 0, –1/2). An orbital's energy, size, shape, or orientation are all unaffected by the electron's spin quantum number (ms). However, it is crucial in determining the electron arrangement within an orbital (Rahm et al., 2019).

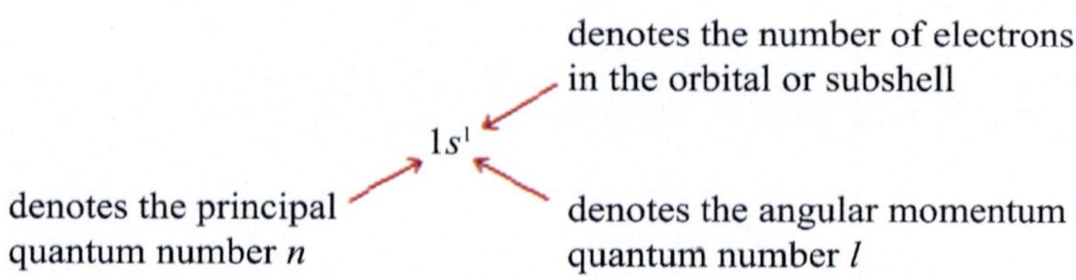

Since the hydrogen atom only has one electron, it is considered a very simple system. The electron can reside in the 1s orbital, also known as the ground state, or it can choose to reside in an orbital with higher energy, also known as an excited state. A thorough grasp of electronic configuration is essential in setting atoms with multiple electrons. (Jin et al., 2020). Understanding this information is crucial for comprehending these atoms' electrical characteristics and behavior. First, the 10 elements of the periodic table will be used, which range from hydrogen to neon, to explain the fundamental concepts that control the arrangement of electron configurations in atoms at their ground state. In this discussion, it is crucial to remember that an atom's atomic number, represented by the letter Z, and the number of electrons it contains are intimately associated (Li et al., 2020).

7.11. THE PAULI EXCLUSION PRINCIPLE

The Pauli Exclusion Principle is a fundamental principle within the field of quantum mechanics that dictates the behavior of particles at both the atomic and subatomic scales. The principle was formulated in 1925 by the renowned Austrian physicist Wolfgang Pauli and was subsequently named after him. The principle above holds significant importance in comprehending substances' composition and has extensive ramifications across diverse scientific disciplines. This section aims to examine the fundamental nature of the Pauli Exclusion Principle, tracing its historical origins and explaining its profound implications within contemporary physics (Milotti et al., 2018).

7.11.1. Historical Background

During the initial decades of the 20th century, physicists were confronted with the nascent discipline of quantum mechanics, which sought to discuss the phenomena exhibited by particles at the microscopic level. The field of quantum mechanics has significantly transformed the comprehension of matter and energy, unveiling notable findings such as wave-particle duality and the quantization of energy.

The Exclusion Principle was introduced by Wolfgang Pauli in 1925 as a resolution to a particular issue within the field of quantum mechanics (Curceanu et al., 2017). During that period, physicists were engaged in the pursuit of comprehending the atomic structure through the explanation of electron distribution surrounding the nucleus.

In accordance with the Bohr model, electrons were postulated to inhabit discrete orbits encircling the nucleus, analogous to the celestial bodies orbiting the sun. Nevertheless, this model proved inadequate in explaining certain phenomena, particularly the emission and absorption of light by atomic systems. The limitations inherent in the Bohr model prompted the development of a more advanced quantum mechanical framework (Shi et al., 2018).

7.11.2. Formulation of the Pauli Exclusion Principle

The fundamental contribution made by Wolfgang Pauli was the recognition that electrons exhibit particle-like behavior and possess an inherent characteristic known as "spin." Spin is a fundamental quantum property that can be conceptualized as the inherent angular momentum of an electron. Pauli observed that electrons possessing half-integer spin are subject to a distinct constraint: it is not permissible for two electrons within an atom to possess identical quantum numbers (Nolte et al., 1990).

Four quantum numbers can each uniquely describe an atom's electron:

1. An important factor that establishes an electron's energy level and roughly estimates its distance from the nucleus is the primary quantum number (n).
2. A crucial element that describes the geometrical arrangement of an electron's orbital is the Angular Momentum Quantum Number (l).
3. One factor that affects how an orbital is oriented in space is its magnetic quantum number (m).
4. An electron's spin is described by the spin quantum number(s), which has two possible values: +1/2 (representing spin-up) or –1/2 (representing spin-down) (Elliott et al., 2012).

The Pauli Exclusion Principle postulates that two electrons within an atom cannot possess identical values for all four quantum numbers. According to this principle, it is postulated that within a specific orbital, a maximum of two electrons is permissible, with one exhibiting a spin-up state (+1/2) and the other displaying a spin-down state (–1/2) (Kaplan, 2020).

7.11.3. Significance and Consequences

The Pauli Exclusion Principle has significant implications for the characteristics of matter, particularly in the comprehension of atomic electronic structure and the periodic table. Determining the physical and chemical characteristics of elements and compounds is critical.

The Pauli Exclusion Principle determines the placement of electrons in atoms. Per the principles of quantum mechanics, electrons are distributed in the lowest available energy levels, with a maximum of two electrons per orbital due to their opposing spins. The principle above holds significant importance in explaining the arrangement of elements within the periodic table and their organization according to their respective electron configurations (Simenel et al., 2017).

The stability of matter, as currently understood, is attributed to the Exclusion Principle. If electrons were to deviate from adhering to this principle and could occupy identical quantum states, the resultant repulsive forces between particles of like charge would be of such magnitude that matter would collapse. This would consequently give rise to a universe characterized by significant instability distinct from the one currently observed (Tennie et al., 2017).

Chemical bonding is a fundamental principle that plays a pivotal role in comprehending the nature of chemical interactions. When atoms combine to create molecules, the interaction between their electrons occurs, and the Pauli Exclusion Principle governs the sharing of electrons between atoms. This phenomenon results in the creation of discrete molecular orbitals and contributes to the overall stability of molecules (Smart et al., 2019).

White dwarf stars are subject to the influence of the Pauli Exclusion Principle, which contributes to comprehending their characteristics. The celestial bodies under consideration are the residual remnants of stars with low to medium masses after the depletion of their nuclear fuel. The exclusion principle dictates that electrons within a white dwarf undergo a process of dense packing, thereby inhibiting any subsequent compression. The phenomenon above gives rise to distinctive characteristics, such as electron degeneracy pressure and a counteractive force against gravitational collapse, thereby establishing stability within the stellar structure (Bartalucci et al., 2006).

Neutron stars, the remains of massive stars following supernova explosions, are subject to the influence of the Pauli Exclusion Principle. Neutron stars exhibit an exceptionally high density, with their central regions comprising tightly packed neutrons. The stability of this dense matter and the emergence of neutron degeneracy pressure are attributed to the governing principle.

In summary, the Pauli Exclusion Principle is a fundamental tenet of quantum mechanics that serves as a pivotal principle dictating the actions of particles on both the atomic and subatomic scales. The phenomenon under consideration has significant implications pertaining to the fundamental composition of matter, the process of chemical bonding, and the characteristics exhibited by celestial entities such as white dwarfs and neutron stars. A thorough grasp of the Pauli Exclusion Principle is crucial for gaining insight into the complexities of the subatomic particles and establishing the fundamental principles that underlie a significant portion of contemporary physics and chemistry (D Javorsek et al., 2000).

7.12. HUND'S RULE

Hund's Rule is a fundamental principle within the field of quantum mechanics that regulates the process of electron orbital filling in atomic systems. The Rule was formulated by Friedrich Hund, a German physicist, during the early 20th century. Hund's Rule principle is paramount in comprehending atoms' electronic configuration and ascertaining their chemical characteristics. This chapter examines the fundamental principles of Hund's Rule, tracing its historical origins and explaining its relevance within contemporary atomic theory (Haule & Kotliar, 2009).

7.12.1. Historical Background

The investigation of electron configuration in atoms can be traced back to the early 20th century when physicists were actively engaged in developing quantum mechanics. The atomic model formulated by Niels Bohr in 1913, commonly referred to as the Bohr model, effectively discussed the line spectra exhibited by hydrogen. However, its explanatory power was limited when explaining the electronic configurations of more complex atoms. The constraints inherent in the Bohr model compelled researchers to pursue a more all-encompassing methodology.

With the advent of quantum mechanics, it became apparent that electrons possess characteristics that can be described in terms of particles and waves. Erwin Schrödinger's formulation of the Schrödinger equation in 1926 established a mathematical framework for explaining the wave-like characteristics of electrons within atomic systems. However, a deeper comprehension of the process by which electrons occupy atomic orbitals necessitated additional inquiry and theoretical advancement (Häkkinen, 2008).

7.12.2. Formulation of Hund's Rule

Friedrich Hund made noteworthy contributions to comprehending atomic structure and the dynamics of electrons within atomic systems. Hund's Rule was introduced by him in 1925 as a significant contribution to the field of atomic spectra and electron configurations.

Hund's Rule postulates that in the presence of multiple degenerate orbitals, electrons prefer to occupy distinct orbitals with identical spin orientations prior to forming pairs. Stated differently, electrons tend to occupy unoccupied orbitals with spins aligned in the same direction before pairing up with spins oriented in the opposite direction.

To discuss Hund's Rule, now contemplate the trio of p-orbitals (px, py, and pz) with equivalent energy within a designated energy level. As per Hund's Rule, electrons will first occupy these three orbitals with parallel spins before any pairing occurs. This results in a configuration wherein there exist unpaired electrons occupying distinct orbitals, which are frequently characterized as possessing "parallel spins" (Colpa et al., 1975).

7.12.3. Significance and Implications

Hund's Rule plays a crucial role in determining atomic electronic configurations, subsequently impacting atoms' chemical properties and behavior. Comprehensive detail of electron configurations is paramount in accurately predicting an atom's reactance, bonding capacities, and spectroscopic characteristics (Parrondo et al., 1993).

1. **Stability of Atoms:** Hund's Rule is a fundamental principle determining the most energetically favorable electronic configurations. Atoms can attain higher stability by minimizing their energy states and maximizing electrons unpaired with parallel spins.
2. **Chemical Bonding:** The chemical bonding behavior of atoms is significantly influenced by their electronic configurations. Atoms that possess unpaired electrons in their valence shells typically exhibit higher reactivity and tend to engage in chemical bonding to attain a more stable electronic configuration.
3. **Paramagnetism:** Hund's Rule also discusses specific magnetic properties. Atoms or ions that possess unpaired electrons demonstrate paramagnetic characteristics, indicating their tendency to be mildly attracted to an external magnetic field.
4. **Periodic Trends:** Hund's Rule plays a significant role in the arrangement of elements within the periodic table. A comprehensive detail of the electron-filling patterns within distinct atomic orbitals facilitates the explication of trends observed in atomic size, energy of ionization, and electronegativity across the periodic table.
5. **Electron Spin Alignment:** The phenomenon of parallel spin alignment, as dictated by Hund's Rule, is a crucial factor in comprehending the magnetic characteristics of substances and how electron spins interact to generate magnetic properties (Andriotis et al., 2004).

7.12.4. Basis of Electron Configuration

The electron configuration refers to the arrangement of electrons in an atom's energy levels and atomic orbitals. The concept holds significant importance within the field of chemistry as it delineates an atom's chemical nature, reactivity, and general stability. The arrangement of electrons in an atom's energy levels results from the atom's electronic structure.

An atom's electrons are arranged in distinct regions known as shells, which correspond to different energy levels. Additionally, these shells are further divided into subshells called atomic orbitals. Principal quantum numbers (n=1, n=2, n=3, etc.) are used to represent shells, whereas subshells are distinguished by the combination of the principal quantum number (n) and the azimuthal quantum number (l). For every given value of l, a total of $(2l + 1)$ orbitals possess distinct orientations.

This can be observed when the quantum number l equals zero, indicating the presence of the s orbital. The s orbital possesses a singular orientation, denoted as 1s.

When the quantum number l equals 1, it corresponds to the p orbital, which exhibits three distinct orientations. The theoretical framework of electron configuration, in conjunction with the Pauli Exclusion Principle and the Aufbau Principle, offers a systematic approach for populating orbitals with electrons by the principles of quantum mechanics (Brik & Srivastava, 2017).

1. **Carbon (Atomic Number 6):** Carbon has an electron configuration of $1s^2\ 2s^2\ 2p^2$. Following Hund's Rule, the two electrons within the 2p subshell are initially distributed into distinct p orbitals before their eventual pairing.

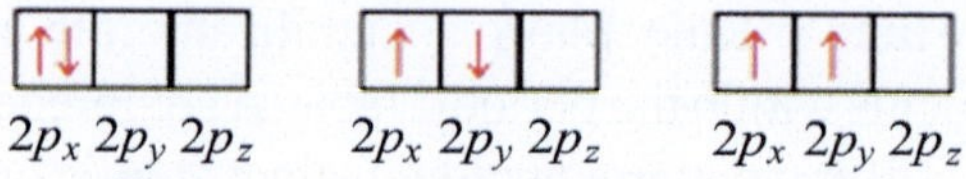

(a) (b) (c)

All three arrangements under consideration adhere to the Pauli exclusion principle. Consequently, the task is to ascertain which arrangement will yield the highest level of stability.

The solution is discussed by Hund's Rule, named after the German physicist Frederick Hund. According to this Rule, the most stable configuration of electrons within subshells is the arrangement that maximizes the number of parallel spins. The configuration depicted in (c) fulfills this criterion. The two spins' cancellation occurs in scenarios (a) and (b). Hence, the orbital diagram representing the electronic configuration of carbon is as follows:

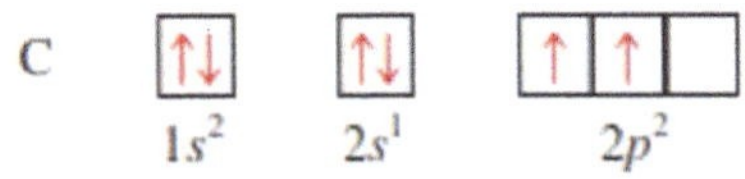

2. **Oxygen (Atomic Number 8):** Oxygen has an electron configuration of $1s^2\ 2s^2\ 2p^4$. In the 2p subshell, the four electrons distribute themselves among the three degenerate p orbitals before any electron pairing (Ranjbar et al., 2011).

3. **Nitrogen (Atomic Number 7):** The electron configuration of nitrogen is represented as $1s^2\ 2s^2\ 2p^3$. In the 2p subshell, the three electrons are distributed individually among the p orbitals before any electron pairing.

4. **Fluorine (Atomic Number 9):** The electron configuration of fluorine, with an atomic number of 9, can be represented as $1s^2\ 2s^2\ 2p^5$. The arrangement of the nine electrons is as follows:

Hund's Rule is a fundamental principle within the field of quantum mechanics that dictates the specific arrangement of electrons within atomic orbitals. Through comprehension of this principle, scientists can interpret the electronic structures of atoms and make projections about their chemical characteristics and actions. Hund's Rule discusses various aspects of atomic stability, chemical bonds, periodic patterns, and magnetic properties. Integrating the concept into contemporary atomic theory has significantly enhanced the comprehension of sub-atomic phenomena. It remains a fundamental pillar in examining electronic configuration and the characteristics of substances on the atomic scale (Gillespie et al., 1996).

Did You Know?

Hund's Rule is an interesting aspect of electron configuration. When multiple orbitals of the same energy level are available, electrons preferentially occupy different orbitals with the same spin before pairing up. This leads to the phenomenon of "parallel spins."

7.13. SUMMARY

The chapter explores the groundbreaking advancements made in the early 20th century that established the fundamental principles for comprehending the characteristics of electrons within atomic systems. This chapter also discusses the fundamental principles of quantum mechanics that govern the electronic configuration of atoms, ranging from electromagnetic radiation to the Pauli Exclusion Principle and Hund's Rule.

The chapter commences by conducting an analysis of electromagnetic radiation, encompassing diverse manifestations of energy propagating in the form of waves. The text introduces Max Planck's seminal Quantum Theory, which postulated the concept of energy quantization in discrete units referred to as "quanta." This theory significantly contributed to the paradigm shift from classical physics to the quantum domain.

One notable aspect of the chapter is the Photoelectric Effect, a phenomenon that perplexed scientists until Albert Einstein reached a sophisticated resolution. The experiment demonstrated that light exhibits dual characteristics, manifesting as wave-like and particle-like entities known as photons. The phenomenon of the photoelectric effect demonstrated that the interaction between light and matter can result in the emission of electrons from a material, provided that the photons possess adequate energy. This observation served as additional evidence supporting the quantum characteristics inherent in light.

Subsequently, the chapter examines Niels Bohr's seminal Theory of the Hydrogen Atom. The model proposed by Bohr posits that electrons are confined to discrete energy levels or orbits encircling the atomic nucleus. The phenomenon of electrons absorbing or emitting energy in distinct quantities during their transitions between energy levels manifests characteristic spectral lines. Although the Bohr model of the atom was a significant advancement in its time, it was not without its constraints. However, its shortcomings ultimately played a crucial role in the subsequent development of quantum mechanics.

The concept of the Dual Nature of the Electron is of significant importance, as it underscores the ability of electrons to manifest characteristics of waves and particles. The electron's wave-like characteristics enable comprehension of phenomena such as interference and diffraction, while its particle-like attributes account for its discrete energy levels and quantized behavior.

The subsequent chapter explores Quantum Mechanics, a groundbreaking theoretical framework that discusses the dynamics of particles at the atomic and subatomic levels. The field of quantum mechanics introduces the fundamental notion of wavefunctions, which are mathematical representations of particles that encompass both wave-like and particle-like characteristics.

Fundamental objects known as quantum numbers are used to describe the state of an electron within an atomic system. The quantum numbers include the Principal Quantum Number (n), used to identify energy levels; the Angular Momentum Quantum Number (l), used to describe the shape of atomic orbitals; the Magnetic Quantum Number (m), used to describe the spatial orientation of orbitals, and the Spin Quantum Number(s), used to describe the spin of electrons.

After establishing the groundwork, the chapter delves into Atomic Orbitals, which refer to specific regions in space that exhibit the highest probability of electron presence. The quantum numbers determine the various shapes and orientations of these orbitals. The shapes above encompass s, p, d, and f orbitals, each exhibiting unique characteristics. Hund's Rule provides insight into how electrons occupy orbitals. Per this principle, electrons will distribute themselves in distinct orbitals within a given energy level, exhibiting identical spin states, before forming pairs with opposite spins.

In summary, the chapter provides a foundation for understanding the fundamental principles that govern the electronic structure of matter and shape our comprehension of the microscopic world, starting from the quantization of energy to the organization of electrons in atomic orbitals.

REFERENCES

1. Agullo, I., Ashtekar, A., & Nelson, W., (2013). Extension of the quantum theory of cosmological perturbations to the Planck era. *Physical Review D, 87*(4), 043507.
2. Aikens, C. M., (2018). Electronic and geometric structure, optical properties, and excited state behavior in atomically precise thiolate-stabilized noble metal nanoclusters. *Accounts of Chemical Research, 51*(12), 3065–3073.
3. Andriotis, A. N., Mpourmpakis, G., Froudakis, G. E., & Menon, M., (2004). Magnetic enhancement and magnetic reduction in binary clusters of transition metal atoms. *The Journal of Chemical Physics, 120*(24), 11901–11904.
4. Aquilanti, V., & Grossi, G., (1980). Angular momentum coupling schemes in the quantum mechanical treatment of P - state atom collisions. *The Journal of Chemical Physics, 73*(3), 1165–1172.
5. Arbó, D. G., Dimitriou, K. I., Persson, E., & Burgdörfer, J., (2008). Sub-Poissonian angular momentum distribution near threshold in atomic ionization by short laser pulses. *Physical Review A, 78*(1), 013406.
6. Auböck, G., Nagl, J., Callegari, C., & Ernst, W. E., (2008). Electron spin pumping of Rb atoms on He nanodroplets via nondestructive optical excitation. *Physical Review Letters, 101*(3), 035301.
7. Ayers, P. W., Morrison, R. C., & Parr, R. G., (2005). Fermi-Amaldi model for exchange-correlation: Atomic excitation energies from orbital energy differences. *Molecular Physics, 103*(15–16), 2061–2072.
8. Bader, R. F., (2005). The quantum mechanical basis of conceptual chemistry. *Monatshefte für Chemie/Chemical Monthly, 136*, 819–854.
9. Barbara, P. F., Walsh, P. K., & Brus, L. E., (1989). Picosecond kinetic and vibrationally resolved spectroscopic studies of intramolecular excited-state hydrogen atom transfer. *The Journal of Physical Chemistry, 93*(1), 29–34.
10. Barnum, D. W., (1961). Electronic absorption spectra of acetylacetonato complexes—II: Hückel LCAO-MO calculations for complexes with trivalent transition metal ions. *Journal of Inorganic and Nuclear Chemistry, 22*(3–4), 183–191.

11. Bartalucci, S., Bertolucci, S., Bragadireanu, M., Cargnelli, M., Catitti, M., Curceanu, C., & VIP Collaboration., (2006). New experimental limit on the Pauli exclusion principle violation by electrons. *Physics Letters B*, *641*(1), 18–22.
12. Bohr, N., (1932). Faraday lecture. Chemistry and the quantum theory of atomic constitution. *Journal of the Chemical Society (Resumed)*, 10(3), 349–384.
13. Bopegedera, A. M. R. P., (2011). A guided-inquiry lab for the analysis of the Balmer series of the hydrogen atomic spectrum. *Journal of Chemical Education*, *88*(1), 77–81.
14. Born, M., (2014). Quantum mechanics. In: *My Life: Recollections of a Nobel Laureate* (Vol. 2, pp. 210–221). Routledge.
15. Brik, M. G., & Srivastava, A. M., (2017). Critical review – a review of the electronic structure and optical properties of ions with d3 electron configuration (V^{2+}, Cr^{3+}, Mn^{4+}, Fe^{5+}) and main related misconceptions. *ECS Journal of Solid State Science and Technology*, *7*(1), R3079.
16. Carrington, T., (1964). Angular momentum distribution and emission spectrum of OH (2Σ+) in the photodissociation of H_2O. *The Journal of Chemical Physics*, *41*(7), 2012–2018.
17. Casati, G., Chirikov, B. V., Shepelyansky, D. L., & Guarneri, I., (1987). Relevance of classical chaos in quantum mechanics: The hydrogen atom in a monochromatic field. *Physics Reports*, *154*(2), 77–123.
18. Celano, U., Goux, L., Belmonte, A., Opsomer, K., Degraeve, R., Detavernier, C., & Vandervorst, W., (2015). Understanding the dual nature of the filament dissolution in conductive bridging devices. *The Journal of Physical Chemistry Letters*, *6*(10), 1919–1924.
19. Clark, T. M., & Chamberlain, J. M., (2014). Use of a PhET interactive simulation in general chemistry laboratory: Models of the hydrogen atom. *Journal of Chemical Education*, *91*(8), 1198–1202.
20. Colpa, J. P., Thakkar, A. J., Smith Jr, V. H., & Randle, P., (1975). An analysis of energy differences in atomic multiplets in connection with the inequality formulation of Hund's rules: I. Frozen versus relaxed SCF orbitals and energies. *Molecular Physics*, *29*(6), 1861–1875.
21. Curceanu, C., Shi, H., Bartalucci, S., Bertolucci, S., Bazzi, M., Berucci, C., & Zmeskal, J., (2017). Test of the Pauli exclusion principle in the VIP-2 underground experiment. *Entropy*, *19*(7), 300.

22. D Javorsek, I. I., Bourgeois, M., Elmore, D., Fischbach, E., Hillegonds, D., Marder, J., & Vogt, S., (2000). New experimental test of the Pauli exclusion principle using accelerator mass spectrometry. *Physical Review Letters*, *85*(13), 2701.

23. Dirac, P. A. M., (1926). Quantum mechanics and a preliminary investigation of the hydrogen atom. *Proceedings of the Royal Society of London. Series A, Containing Papers of a Mathematical and Physical Character*, *110*(755), 561–579.

24. Eichler, J., & Stöhlker, T., (2007). Radiative electron capture in relativistic ion–atom collisions and the photoelectric effect in hydrogen-like high-Z systems. *Physics Reports*, *439*(1–2), 1–99.

25. Elliott, S. R., LaRoque, B. H., Gehman, V. M., Kidd, M. F., & Chen, M., (2012). An improved limit on Pauli-exclusion-principle forbidden atomic transitions. *Foundations of Physics*, *42*(8), 1015–1030.

26. Freed, K. F., (1971). Many-body theories of the electronic structure of atoms and molecules. *Annual Review of Physical Chemistry*, *22*(1), 313–346.

27. Fujita, I., Hatada, M., Ogawa, T., & Hirota, K., (1971). The emission spectra of hydrogen atoms produced from methanol by electron impact. *Bulletin of the Chemical Society of Japan*, *44*(7), 1751–1754.

28. Gaeta, Z. D., & Stroud, Jr. C. R., (1990). Classical and quantum-mechanical dynamics of a quasiclassical state of the hydrogen atom. *Physical Review A*, *42*(11), 6308.

29. Gálvez, J. R., Wolf, C., Delgado, F., & Lorente, N., (2019). Cotunneling mechanism for all-electrical electron spin resonance of single adsorbed atoms. *Physical Review B*, *100*(3), 035411.

30. Gillespie, R. J., Spencer, J. N., & Moog, R. S., (1996). Electron configurations from experiment. *Journal of Chemical Education*, *73*(7), 617.

31. Gretarsson, H., Lupascu, A., Kim, J., Casa, D., Gog, T., Wu, W., & Kim, Y. J., (2011). Revealing the dual nature of magnetism in iron pnictides and iron chalcogenides using X-ray emission spectroscopy. *Physical Review B*, *84*(10), 100509.

32. Häkkinen, H., (2008). Atomic and electronic structure of gold clusters: understanding flakes, cages, and superatoms from simple concepts. *Chemical Society Reviews*, *37*(9), 1847–1859.

33. Hall, G. G., (1959). Applications of quantum mechanics in theoretical chemistry. *Reports on Progress in Physics*, *22*(1), 1.
34. Harth, K., Ruf, M. W., & Hotop, H., (1989). Electron transfer from laser excited Rydberg atoms to molecules. Absolute rate constants at low and intermediate principal quantum numbers. *Zeitschrift für Physik D Atoms, Molecules, and Clusters*, *14*, 149–165.
35. Haule, K., & Kotliar, G., (2009). Coherence–incoherence crossover in the normal state of iron oxypnictides and importance of Hund's rule coupling. *New Journal of Physics*, *11*(2), 025021.
36. Herschbach, D., (2001). Molecular beams entwined with quantum theory: A bouquet for Max Planck. *Annalen der Physik*, *513*(1–2), 163–176.
37. Hinton, J. F., Guthrie, P., Pulay, P., & Wolinski, K., (1992). Ab initio quantum mechanical calculation of the chemical shift anisotropy of the hydrogen atom in the (H_2O) 17 water cluster. *Journal of the American Chemical Society*, *114*(5), 1604–1605.
38. Howell, J. M., Absar, I., & Van Wazer, J. R., (1973). Role of sulfur d orbitals in isothiocyanic acid and an electronic - structure comparison with isocyanic acid. *The Journal of Chemical Physics*, *59*(11), 5895–5902.
39. Jin, X., Wang, R., Zhang, L., Si, R., Shen, M., Wang, M., & Shi, J., (2020). Electron configuration modulation of nickel single atoms for elevated photocatalytic hydrogen evolution. *Angewandte Chemie*, *132*(17), 6894–6898.
40. Kaplan, I. G., (2020). The Pauli exclusion principle and the problems of its experimental verification. *Symmetry*, *12*(2), 320.
41. Kelly, H. P., & Carter, S. L., (1980). Many body perturbation calculations of the interaction of atoms with electromagnetic radiation. *Physica Scripta*, *21*(3–4), 448.
42. Kim, J., Baczewski, A. D., Beaudet, T. D., Benali, A., Bennett, M. C., Berrill, M. A., & Zhao, L., (2018). QMCPACK: An open-source ab initio quantum Monte Carlo package for the electronic structure of atoms, molecules, and solids. *Journal of Physics: Condensed Matter*, *30*(19), 195901.
43. Knizia, G., (2013). Intrinsic atomic orbitals: An unbiased bridge between quantum theory and chemical concepts. *Journal of Chemical Theory and Computation*, *9*(11), 4834–4843.

44. Kolbinger, B., Amsler, C., Cuendis, S. A., Breuker, H., Capon, A., Costantini, G., & Zmeskal, J., (2021). Measurement of the principal quantum number distribution in a beam of antihydrogen atoms. *The European Physical Journal D*, *75*, 1–14.

45. Kraft, T., Ruf, M. W., & Hotop, H., (1990). Strong dependence of negative cluster ion spectra on principal quantum number n in collisions of state-selected Ar**(nd) Rydberg atoms with N_2O clusters. *Zeitschrift für Physik D Atoms, Molecules, and Clusters*, *17*, 37–43.

46. Krems, R. V., Groenenboom, G. C., & Dalgarno, A., (2004). Electronic interaction anisotropy between atoms in arbitrary angular momentum states. *The Journal of Physical Chemistry A*, *108*(41), 8941–8948.

47. Kutzelnigg, W., (2008). The principle-quantum-number (and the radial-quantum-number) expansion of the correlation energy of two-electron atoms. *Physical Chemistry Chemical Physics*, *10*(23), 3460–3468.

48. Li, Y., Xu, X., Wang, J., Luo, W., Zhang, Z., Cheng, X., & Wang, L., (2020). Post-redox engineering electron configurations of atomic thick C_3N_4 nanosheets for enhanced photocatalytic hydrogen evolution. *Applied Catalysis B: Environmental*, *270*, 118855.

49. Marcus, R. A., (1966). On the analytical mechanics of chemical reactions. Quantum mechanics of linear collisions. *The Journal of Chemical Physics*, *45*(12), 4493–4499.

50. Mills, R. L., (2000). The hydrogen atom revisited. *International Journal of Hydrogen Energy*, *25*(12), 1171–1183.

51. Milotti, E., Bartalucci, S., Bertolucci, S., Bazzi, M., Bragadireanu, M., Cargnelli, M., & Zmeskal, J., (2018). On the importance of electron diffusion in a bulk-matter test of the Pauli exclusion principle. *Entropy*, *20*(7), 515.

52. Mulliken, R. S., (1932). The interpretation of band spectra part III. Electron quantum numbers and states of molecules and their atoms. *Reviews of Modern Physics*, *4*(1), 1.

53. Nakiboglu, C., (2003). Instructional misconceptions of Turkish prospective chemistry teachers about atomic orbitals and hybridization. *Chemistry Education Research and Practice*, *4*(2), 171–188.

54. Neese, F., Atanasov, M., Bistoni, G., Maganas, D., & Ye, S., (2019). Chemistry and quantum mechanics in 2019: Give us insight and numbers. *Journal of the American Chemical Society*, *141*(7), 2814–2824.

55. Nolte, E., Faestermann, T., Gail, H., Gillitzer, A., Korschinek, G., Müller, D., & Logan, B. A., (1990). Search for a violation of the Pauli exclusion principle for atomic electrons and for nucleons. *Nuclear Instruments and Methods in Physics Research Section B: Beam Interactions with Materials and Atoms*, *52*(3–4), 563–567.
56. Ossiander, M., Riemensberger, J., Neppl, S., Mittermair, M., Schäffer, M., Duensing, A., & Kienberger, R., (2018). Absolute timing of the photoelectric effect. *Nature*, *561*(7723), 374–377.
57. Ozaki, T., (2003). Variationally optimized atomic orbitals for large-scale electronic structures. *Physical Review B*, *67*(15), 155108.
58. Parrondo, R. M., Karafiloglou, P., & Marcos, E. Ś., (1993). Examination of the Hund rule in closed - shell systems: Investigation of spin correlation effects. *International Journal of Quantum Chemistry*, *47*(3), 191–211.
59. Pauling, L. C., (1927). The theoretical prediction of the physical properties of many electron atoms and ions. Mole refraction, diamagnetic susceptibility, and extension in space. *Proceedings of the Royal Society of London. Series A, Containing Papers of a Mathematical and Physical Character*, *114*(767), 181–211.
60. Picozzi, S., Yamauchi, K., Sanyal, B., Sergienko, I. A., & Dagotto, E., (2007). Dual nature of improper ferroelectricity in a magnetoelectric multiferroic. *Physical Review Letters*, *99*(22), 227201.
61. Pratt, R. H., (1960). Atomic photoelectric effect at high energies. *Physical Review*, *117*(4), 1017.
62. Raffenetti, R. C., (1973). Even - tempered atomic orbitals. II. Atomic SCF wavefunctions in terms of even - tempered exponential bases. *The Journal of Chemical Physics*, *59*(11), 5936–5949.
63. Rahm, M., Cammi, R., Ashcroft, N. W., & Hoffmann, R., (2019). Squeezing all elements in the periodic table: Electron configuration and electronegativity of the atoms under compression. *Journal of the American Chemical Society*, *141*(26), 10253–10271.
64. Ranjbar, A., Babamoradi, M., Saani, M. H., Vesaghi, M. A., Esfarjani, K., & Kawazoe, Y., (2011). Many-electron states of nitrogen-vacancy centers in diamond and spin density calculations. *Physical Review B*, *84*(16), 165212.
65. Richardson, J. W., Nieuwpoort, W. C., Powell, R. R., & Edgell, W. F., (1962). Approximate radial functions for first - row transition - metal

atoms and ions. I. Inner - shell, 3d and 4s atomic orbitals. *The Journal of Chemical Physics*, *36*(4), 1057–1061.

66. Sakimoto, K., (2001). Full quantum-mechanical study of protonium formation in slow collisions of antiprotons with hydrogen atoms. *Physical Review A*, *65*(1), 012706.
67. Schneider, B. I., Hamilton, K. R., & Bartschat, K., (2022). Generalizations of the R-matrix method to the treatment of the interaction of short-pulse electromagnetic radiation with atoms. *Atoms*, *10*(1), 26.
68. Senet, P., (1997). Chemical hardnesses of atoms and molecules from frontier orbitals. *Chemical Physics Letters*, *275*(5–6), 527–532.
69. Shen, C., Li, X., Wang, T., Wang, G., & Xia, C., (2022). Effects of p–d orbital mixings on magnetocrystalline anisotropy for D 3d-symmetric ferromagnetic semiconducting monolayers. *Physical Chemistry Chemical Physics*, *24*(39), 24553–24561.
70. Shi, H., Milotti, E., Bartalucci, S., Bazzi, M., Bertolucci, S., Bragadireanu, A. M., & Curceanu, C., (2018). Experimental search for the violation of Pauli exclusion principle: VIP-2 collaboration. *The European Physical Journal C*, *78*, 1–14.
71. Shields, G. C., & Kash, M. M., (1992). Experiment in quantization: Atomic line spectra. *Journal of Chemical Education*, *69*(4), 329.
72. Shirley, D. A., (1972). The effect of atomic and extra-atomic relaxation on atomic binding energies. *Chemical Physics Letters*, *16*(2), 220–225.
73. Simenel, C., Umar, A. S., Godbey, K., Dasgupta, M., & Hinde, D. J., (2017). How the Pauli exclusion principle affects fusion of atomic nuclei. *Physical Review C*, *95*(3), 031601.
74. Singh, G., (2012). Computer simulations of quantum theory of hydrogen atom for natural science education students in a virtual lab. *Journal of Educational Technology Systems*, *40*(3), 273–286.
75. Singleton, S. M., Teague, C. M., & Salter, C., (2022). The hydrogen atom spectrum: Experimental analysis using iterative model building. *Journal of Chemical Education*, *99*(12), 4143–4148.
76. Smart, S. E., Schuster, D. I., & Mazziotti, D. A., (2019). Experimental data from a quantum computer verifies the generalized Pauli exclusion principle. *Communications Physics*, *2*(1), 11.
77. Sokolov, I. M., Kupriyanov, D. V., & Havey, M. D., (2011). Microscopic theory of scattering of weak electromagnetic radiation by a dense

ensemble of ultracold atoms. *Journal of Experimental and Theoretical Physics*, *112*, 246–260.

78. Stamplecoskie, K. G., Manser, J. S., & Kamat, P. V., (2015). Dual nature of the excited state in organic–inorganic lead halide perovskites. *Energy & Environmental Science*, *8*(1), 208–215.
79. Suto, K., (2021). The quantum condition that should have been assumed by Bohr when deriving the energy levels of a hydrogen atom. *Journal of Applied Mathematics and Physics*, *9*(6), 1230–1244.
80. Tennie, F., Vedral, V., & Schilling, C., (2017). Influence of the fermionic exchange symmetry beyond Pauli's exclusion principle. *Physical Review A*, *95*(2), 022336.
81. Thakkar, A. J., (2003). The momentum density perspective of the electronic structure of atoms and molecules. *Advances in Chemical Physics*, *128*, 303–352.
82. Tsaparlis, G., (1997). Atomic orbitals, molecular orbitals and related concepts: Conceptual difficulties among chemistry students. *Research in Science Education*, *27*, 271–287.
83. Vander Sluis, K. L., & Nugent, L. J., (1974). Systematics in the relative energies of some low-lying electron configurations in the gaseous atoms and free ions of the lanthanide and actinide series. *JOSA*, *64*(5), 687–695.
84. Verner, E. M., Verner, D. A., Korista, K. T., Ferguson, J. W., Hamann, F., & Ferland, G. J., (1999). Numerical simulations of Fe II emission spectra. *The Astrophysical Journal Supplement Series*, *120*(1), 101.
85. Yuan, H., Yang, Y., Guo, F., Wang, J., Chen, J., Feng, W., & Cui, Z., (2023). Effect of pulse duration on the above-threshold ionization of a hydrogen atom irradiated by a 400 nm intense laser. *Optics Express*, *31*(15), 24213–24229.
86. Zhao, Z., Cao, Y., Yi, J., He, X., Ma, C., & Qiu, J., (2012). Band - edge electronic structure of β - In_2S_3: The role of s or p orbitals of atoms at different lattice positions. *ChemPhysChem*, *13*(6), 1551–1556.
87. Zhou, Z., & Yuan, J., (2008). Fine structures of the harmonic and hyper-Raman spectrum of the hydrogen atom in an intense high-frequency laser pulse. *Physical Review A*, *77*(6), 063411.

CHAPTER 8

Modern Techniques for Studying Atoms

Contents

8.1. INTRODUCTION

This chapter discusses the modern techniques for studying atoms. These cutting-edge methodologies have transformed the understanding of atomic structures, elemental compositions, and interactions from nanoscale imaging to precise mass measurements and spectroscopic analyses. Gone are the days of crude models and indirect evidence; humans now possess the tools to peer directly into the atom's heart and explore its mysteries (Urban, 2008).

Among the remarkable techniques, scanning tunneling microscopy (STM) is a pioneer in nanoscale visualization. With the ability to resolve individual atoms and manipulate them with astonishing precision, STM has revolutionized the understanding of surfaces and materials, opening the door to nanotechnology and innovative applications in various industries (Somorjai, 1996).

Accompanying STM, atomic force microscopy (AFM) offers complementary insights into nanoscale forces and mechanical properties. Employing various imaging modes, AFM allows scientists to map surfaces, detect forces, and explore the biological field with unprecedented detail.

The exploration does not end there; Transmission Electron Microscopy (TEM) allows us to venture even deeper into the atomic world, providing high-resolution images of atomic structures and crystal defects. Coupled with electron diffraction, TEM has facilitated breakthroughs in materials science, enabling to engineer and analysis of materials on the atomic scale.

Moving beyond imaging, X-ray Crystallography offers a profound understanding of crystal structures, unraveling the arrangement of atoms in three-dimensional space. This indispensable technique has revealed the secrets of biomolecules, inorganic compounds, and solid-state materials, propelling advancements in chemistry and biology (Tanner et al., 2000).

Mass Spectrometry, another revolutionary technique, empowers us to measure the masses of atoms and molecules with extraordinary precision. This method not only aids in elemental identification but also unlocks a wealth of information about isotopes, molecular structures, and complex mixtures.

In nuclear magnetic resonance (NMR) spectroscopy, the magnetic properties of atomic nuclei unveil intricate molecular structures, providing valuable insights into organic compounds, proteins, and complex molecules. Through NMR, researchers are at liberty to unravel the secrets of biomolecular interactions and pave the way for novel pharmaceutical developments (Kuznetsova et al., 2007). With its precision and versatility, Laser Spectroscopy embraces the world of atoms and molecules, enabling researchers to probe their electronic and vibrational transitions. This powerful technique has applications in atmospheric studies, analytical chemistry, and fundamental physics research. As we dive deeper into Time-of-Flight Mass Spectrometry (TOF-MS), we uncover a technique capable of rapid, high-resolution mass measurements, accelerating the pace of proteomics, metabolomics, and petroleomics. TOF-MS is pushing the boundaries of analytical chemistry, unleashing new dimensions of information from complex samples (Nandi & Ajnavarapu, 2021). This chapter celebrates the ingenuity and dedication of scientists who have harnessed these modern techniques, propelling the understanding of atoms to unprecedented heights.

Tip

X-ray crystallography played a pivotal role in discovering the structure of DNA, leading to the famous double helix model proposed by Watson and Crick in 1953.

8.2. SCANNING TUNNELING MICROSCOPY (STM)

In nanoscience and nanotechnology, the ability to explore and manipulate matter at the atomic scale has been a transformative force. Among the remarkable tools that enable such feats, scanning tunneling microscopy (STM) is a pioneering technique that has revolutionized the understanding of the atomic world.

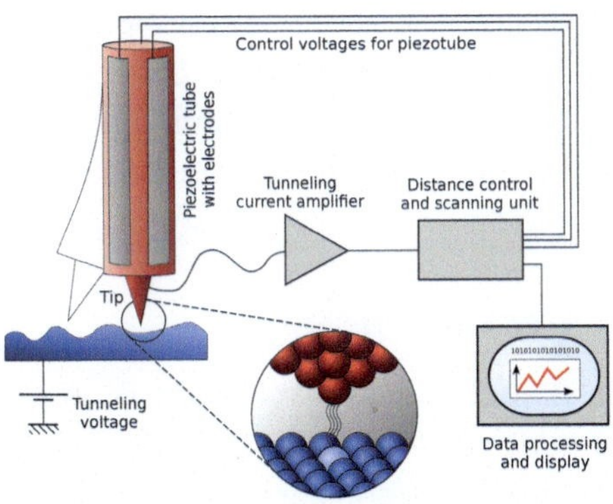

Figure 8.1. *Principle of scanning electron microscopy.*

Source: Bardeen, Creative Commons License.

Developed in the early 1980s by Gerd Binnig and Heinrich Rohrer, STM opened a window into the sub-atomic world of individual atoms, revealing their spatial distribution and electronic properties with astonishing precision (Tersoff & Hamann, 1985) (Figure 8.1).

8.2.1. The Principles of Scanning Tunneling Microscopy (STM)

8.2.1.1. Quantum Tunneling Phenomenon

The foundation of STM lies in the quantum mechanical phenomenon known as tunneling. When a conducting probe is brought very close to a conducting surface, such as a metal or a semiconductor, the outermost electrons of both the probe and the surface experience a quantum mechanical effect. As a result, electrons can tunnel through the potential barrier despite not possessing sufficient energy to overcome it in classical terms.

This tunneling current, exponentially dependent on the distance between the probe and the surface, forms the basis for imaging and probing atomic structures (Kuk & Silverman, 1989).

8.2.1.2. Tunneling Current and Distance Dependence

The tunneling current, I, between the tip of the STM probe and the surface is given by the famous expression known as the Fowler-Nordheim equation:

$$I \infty V \times e^{-2k \times d}$$

where; 'V' is the voltage applied between the tip and the surface; 'k' is a constant that depends on the materials' properties; and 'd' is the distance between the tip and the surface. The tunneling current is highly sensitive to the distance between the probe and the surface, making STM an exquisitely precise imaging tool at the atomic level (Tersoff & Hamann, 1983).

8.2.1.3. Feedback Mechanism

STM utilizes a feedback mechanism to maintain a constant tunneling current and keep the tip at a consistent distance from the surface. As the tip scans across the surface, it maintains a constant tunneling current by adjusting the height of the tip using piezoelectric actuators. The feedback loop ensures that the tunneling current remains stable, producing atomic-resolution topographic images (Binnig & Rohrer, 2000).

8.2.2. Instrumentation of Scanning Tunneling Microscopy (STM)

8.2.2.1. The STM Probe

The sharp metallic tip, typically made of tungsten or platinum-iridium, is at the heart of the STM. This tip is brought into proximity with the surface of interest, allowing for the quantum tunneling of electrons. The apex of the tip is ideally terminated with a single atom, conferring exceptional spatial resolution to the STM.

8.2.2.2. Piezoelectric Scanners

To scan the tip across the surface, piezoelectric scanners are employed. These scanners utilize the piezoelectric effect, where certain materials expand or contract in response to an applied electric field. By applying voltage to these scanners, the tip can be precisely moved in the x, y, and z directions, enabling detailed surface topography imaging (Binnig & Rohrer, 1985).

8.2.2.3. Tunneling Current Amplifier and Feedback System

The tunneling current is extremely weak, ranging from a few pico-amps to femto-amps, making it challenging to detect directly. Tunneling current amplifiers magnify the current signal to a measurable range. The amplified current is then fed into a feedback system that adjusts the tip-sample distance to maintain a constant tunneling current during scanning (Hansma & Tersoff, 1987).

8.2.2.4. Electronic Control System

A sophisticated electronic control system coordinates the piezoelectric scanners' movement, the tunneling current measurement, and the feedback loop. The control system also interfaces with a computer, which processes the acquired data to generate high-resolution images.

8.2.3. Scanning Modes in STM

STM operates in different scanning modes, each catering to specific imaging needs. The three primary scanning modes are discussed in subsections.

8.2.3.1. Constant Height Mode

In this mode, the tip is held at a fixed height above the surface during scanning. As the tip moves over surface features, the tunneling current varies due to changes in the local surface height. Constant height mode is useful for capturing real-time dynamic processes, such as surface reactions or diffusion of atoms (Binnig & Rohrer, 2000).

8.2.3.2. Constant Current Mode

The STM feedback loop maintains a constant tunneling current in this mode as the tip scans the surface. The variations in tip height during scanning yield information about the surface topography, providing high-resolution images of atomic structures.

8.2.3.3. Constant Force Mode

Constant force mode maintains a constant interaction force between the tip and the surface. This mode is used when imaging soft or delicate materials, ensuring the tip does not exert excessive force and damage the sample.

8.2.4. Applications of Scanning Tunneling Microscopy (STM)

8.2.4.1. Surface Topography and Atomic Imaging

One of the primary applications of STM is imaging surface topography with atomic resolution. STM has provided breathtaking visualizations of individual atoms on various materials, including metals, semiconductors, and insulators. These images have contributed to the understanding of atomic arrangements, defects, and surface reconstructions (Binnig & Rohrer, 1985).

8.2.4.2. Manipulation of Atoms and Molecules

Beyond imaging, STM allows for the manipulation of individual atoms and molecules. By precisely positioning the tip, scientists can pick up and move atoms or molecules on surfaces, opening the doors to atom-by-atom construction and nanoscale engineering.

8.2.4.3. Surface Science and Catalysis Studies

STM plays a crucial role in surface science, where the interactions between atoms and molecules on surfaces are studied. It has been invaluable in investigating catalysis processes and understanding the dynamics of surface reactions, providing insights into chemical reactions at the atomic level.

8.2.4.4. Quantum Phenomena and Spintronics

STM has also been instrumental in exploring quantum phenomena, such as quantum confinement and electron quantization in nanostructures. Additionally, STM has contributed to the emerging field of spintronics, where the spin of electrons is exploited for information processing and storage (Grafstrom, 2002).

8.2.4.5. Biological Applications

In recent years, STM has found applications in biology and biophysics. It has been used to study biomolecules, such as DNA and proteins, at the single-molecule level, providing insights into their structures and interactions.

Scanning tunneling microscopy (STM) is a crowning achievement in understanding and manipulating matter at the atomic scale. Its exceptional resolution and versatility have empowered scientists across various disciplines to explore the wonders of the nanoscale world. From the precise imaging of individual atoms to manipulating matter atom-by-atom, STM continues to shape our understanding of fundamental physical, chemical, and biological processes. As technology advances, STM will undoubtedly play an even more transformative role in advancing the understanding of the atomic landscape and driving innovations in nanoscience and nanotechnology (Binnig et al., 1982).

Remember

Laser cooling techniques, which use lasers to slow down and cool atoms to near absolute zero, have led to the creation of Bose-Einstein condensates, a new state of matter, and won the Nobel Prize in Physics in 1997.

8.3. ATOMIC FORCE MICROSCOPY (AFM)

Atomic force microscopy (AFM) is a powerful and versatile imaging technique used to investigate the surfaces of materials at the nanoscale. Unlike STM, which relies on the quantum tunneling of electrons, AFM operates on the principle of mechanical interaction between a sharp tip and the sample surface. This interaction allows researchers to achieve exceptionally high-resolution images of atomic and molecular structures and probe surface forces and mechanical properties with incredible precision (Hansma et al., 1988).

In AFM, a sharp tip attached to a flexible cantilever is brought closer to the sample surface. As the tip scans across the surface, the interaction forces between the tip and the atoms or molecules on the surface cause the cantilever to deflect. This deflection is measured by a laser beam reflected off the cantilever, allowing researchers to create a three-dimensional surface topography map with atomic-scale resolution. AFM can operate in various imaging modes, such as contact, tapping, and non-contact modes, each offering unique advantages for specific imaging and measurement needs (Giessibl, 2003) (Figure 8.2).

One of the significant advantages of AFM is its ability to image a wide range of samples, including conductive, insulating, and biological materials. Unlike other imaging techniques like electron microscopy, AFM can be used under ambient conditions, making it well-suited for studying delicate biological samples that might be adversely affected by high vacuum or electron beam damage.

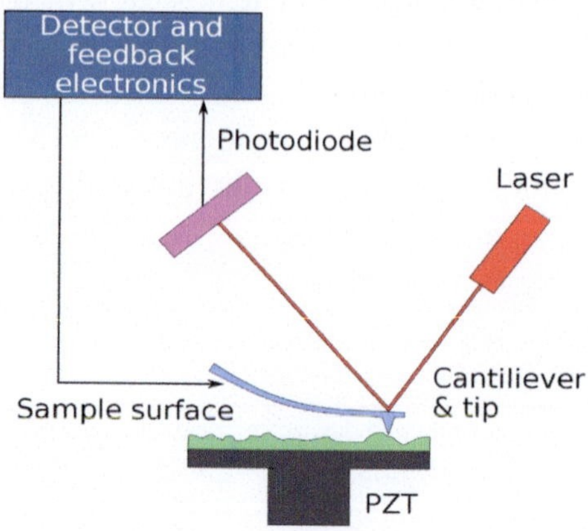

Figure 8.2. *Principle of atomic force microscopy.*

Source: Tersoff, Creative Commons License.

AFM also provides valuable information about the mechanical properties of surfaces. Researchers can gain insights into material properties like elasticity, stiffness, and hardness by measuring the forces required to deform the surface or the adhesion forces between the tip and the sample. This capability is especially useful in materials science, nanomechanics, and biological research, where understanding surface mechanical properties is essential (Garcia & Perez, 2002).

In addition to static imaging, AFM can study dynamic processes in real-time. With high-speed scanning and force measurements, researchers can observe dynamic phenomena like molecular reactions, self-assembly processes, and the mechanical behavior of biological molecules. This real-time capability is particularly advantageous in studying time-dependent processes crucial in various scientific investigations.

The versatility of AFM extends beyond imaging and mechanical probing. AFM-based techniques, such as scanning capacitance microscopy and Kelvin probe force microscopy, enable the measurement of local electrical and electronic properties, making AFM a valuable tool in studying semiconductors and nanostructures. Furthermore, AFM can be coupled with other techniques like Raman spectroscopy, infrared spectroscopy, and fluorescence microscopy, expanding its capabilities to gather more comprehensive information about samples (Rugar & Hansma, 1990).

AFM has become a cornerstone in nanoscale investigations, offering remarkable capabilities to explore and understand surfaces at the atomic and molecular levels. Its non-destructive and versatile nature allows researchers to study various materials and dynamic processes, from solid-state physics and materials science to biological research. AFM continues to play a pivotal role in advancing our understanding of nanoscale phenomena, providing essential insights into the atomic world and driving innovations in diverse scientific disciplines (Cappella & Dietler, 1999).

Did You Know?

The smallest logo ever created was "IBM" using 35 individual xenon atoms, achieved with a scanning tunneling microscope (STM) in 1990.

8.4. TRANSMISSION ELECTRON MICROSCOPY

Transmission Electron Microscopy (TEM) is a powerful imaging technique that allows scientists to study the internal structure of materials with unparalleled resolution. A high-energy electron beam in TEM is transmitted through a thin specimen, revealing details at the atomic and nanoscale levels. This technique has revolutionized our understanding of materials science, solid-state physics, nanotechnology, and biological research, providing crucial insights into the fundamental properties of diverse materials.

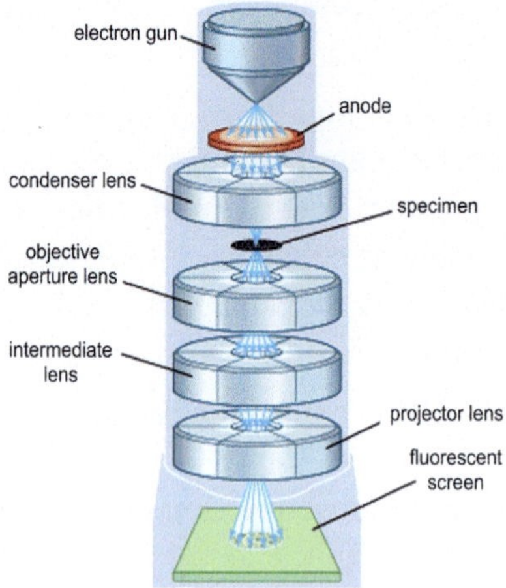

Figure 8.3. *Principle of electron microscopy.*

Source: Hansma, Creative Commons License.

The key components of a TEM setup include an electron source, electromagnetic lenses, a specimen holder, and a detector. The electron source generates a beam of electrons with high energy and narrow wavelength, typically using a heated tungsten filament or a field-emission gun. Electromagnetic lenses focus and direct the electron beam through the specimen while a detector captures the transmitted electrons to generate an image (Meyer, 1992) (Figure 8.3).

One of the unique advantages of TEM is its ability to study thin sections of materials, typically in the range of tens to hundreds of nanometers. Thin specimens are essential for efficient electron transmission and to avoid electron beam scattering. To prepare the samples, researchers use techniques like ultramicrotomy, ion milling, or focused ion beam (FIB) milling to produce thin sections suitable for TEM imaging (Winey et al., 2014).

In addition to revealing atomic structures, TEM is instrumental in studying the interactions between electrons and materials. Electron diffraction patterns obtained from the transmitted electron beam provide information about the crystallographic orientation and symmetry of the specimen.

This diffraction data is used to determine the arrangement of atoms in a crystal lattice and to study the crystallographic defects and grain boundaries in materials (Schorb et al., 2019).

TEM has found widespread applications in various scientific disciplines. In materials science and nanotechnology, TEM allows researchers to characterize the morphology and properties of nanomaterials, such as nanoparticles, nanowires, and thin films. In solid-state physics, TEM is used to investigate the electronic properties of materials, such as band structures and energy band gaps. Moreover, TEM is invaluable in biology for visualizing cellular organelles, viruses, and macromolecular complexes, enabling a deeper understanding of biological processes at the nanoscale.

However, TEM does have some limitations. It requires skilled sample preparation to produce thin sections, and the electron beam can cause damage to sensitive biological samples. Additionally, the high vacuum environment in TEM can affect the behavior of some materials, making them less suitable for studying samples in their native environments (Kohl & Reimer, 2008).

In conclusion, Transmission Electron Microscopy (TEM) is a remarkable tool that has transformed the exploration of the nanoscale world. Its extraordinary spatial resolution and versatility have allowed scientists to uncover atomic structures, study material properties, and visualize nanoscale phenomena in various materials. As TEM technology advances, it will remain a cornerstone of scientific research, driving innovations and discoveries across various fields, from materials science to biology.

Remember

The world's first scanning tunneling microscope (STM) was invented in 1981 by Gerd Binnig and Heinrich Rohrer, leading to their Nobel Prize in Physics in 1986.

8.5. SCANNING ELECTRON MICROSCOPY

Transmission Electron Microscopy (TEM) is a powerful imaging technique that allows scientists to study the internal structure of materials with unparalleled resolution. A high-energy electron beam in TEM is transmitted through a thin specimen, revealing details at the atomic and nanoscale levels. This technique has revolutionized our understanding of materials science, solid-state physics, nanotechnology, and biological research, providing crucial insights into the fundamental properties of diverse materials (Spurgeon et al., 2021).

The key components of a TEM setup include an electron source, electromagnetic lenses, a specimen holder, and a detector. The electron source generates a beam of electrons with high energy and narrow wavelength, typically using a heated tungsten filament or a field-emission gun. Electromagnetic lenses focus and direct the electron beam through the specimen while a detector captures the transmitted electrons to generate an image.

TEM offers exceptional spatial resolution, revealing lattice structures, crystal defects, and individual atoms within a material. By carefully controlling the electron beam's wavelength and focusing, TEM can achieve resolutions down to a fraction of an angstrom, allowing researchers to visualize even the smallest features in materials (Leamy, 1982). One of the unique advantages of TEM is its ability to study thin sections of materials, typically in the range of tens to hundreds of nanometers. Thin specimens are essential for efficient electron transmission and to avoid electron beam scattering. To prepare the samples, researchers use techniques like ultramicrotomy, ion milling, or focused ion beam (FIB) milling to produce thin sections suitable for TEM imaging (Vernon, 2000).

In addition to revealing atomic structures, TEM is instrumental in studying the interactions between electrons and materials. Electron diffraction patterns obtained from the transmitted electron beam provide information about the crystallographic orientation and symmetry of the specimen. This diffraction data is used to determine the arrangement of atoms in a crystal lattice and to study the crystallographic defects and grain boundaries in materials.

TEM has found widespread applications in various scientific disciplines. In materials science and nanotechnology, TEM allows researchers to characterize the morphology and properties of nanomaterials, such as nanoparticles, nanowires, and thin films. In solid-state physics, TEM is used to investigate the electronic properties of materials, such as band structures and energy band gaps. Moreover, TEM is invaluable in biology for visualizing cellular organelles, viruses, and macromolecular complexes, enabling a deeper understanding of biological processes at the nanoscale.

However, TEM does have some limitations. It requires skilled sample preparation to produce thin sections, and the electron beam can cause damage to sensitive biological samples. Additionally, the high vacuum environment in TEM can affect the behavior of some materials, making them less suitable for studying samples in their native environments (Mohammed & Abdullah, 2018).

Transmission Electron Microscopy (TEM) is a remarkable tool that has transformed our ability to explore the nanoscale world. Its extraordinary spatial resolution and versatility have allowed scientists to uncover atomic structures, study material properties, and visualize nanoscale phenomena in various materials. As TEM technology advances, it will remain a cornerstone of scientific research, driving innovations and discoveries across various fields, from materials science to biology.

Remember

Scanning Electron Microscopy (SEM) can provide stunning and artistic images of various materials, including insects, plants, and even food items, due to its nanometer resolution.

8.6. X-RAY CRYSTALLOGRAPHY

X-ray crystallography is a powerful and widely used technique in science that allows researchers to determine crystals' atomic and molecular structure. It is a fundamental tool in chemistry, physics, biology, materials science, and many other scientific disciplines. The technique relies on the principle that when X-rays are directed at a crystal, they are diffracted by the atoms within the crystal lattice, producing a unique pattern of diffraction spots. Analyzing these diffraction patterns provides critical information about the arrangement of atoms within the crystal and allows scientists to reconstruct the three-dimensional structure of the crystal (Scheinfein et al., 1990).

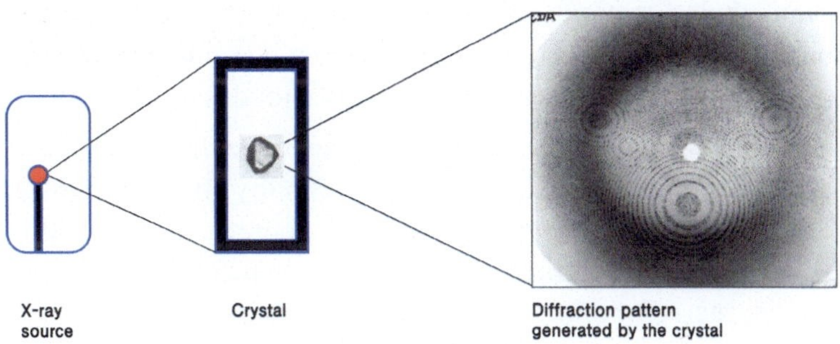

Figure 8.4. *Principle of X-ray crystallography.*

Source: Brian, Creative Commons License.

The history of X-ray crystallography can be traced back to the early 20th century when Bragg's law was formulated by William Henry Bragg and his son William Lawrence Bragg. Bragg's law relates the angles of diffraction to the spacing of the crystal lattice planes, providing a fundamental basis for crystallography. The father-son duo's pioneering work earned them the Nobel Prize in Physics in 1915, and it marked the beginning of a new era in the understanding of crystal structures (Figure 8.4).

In modern X-ray crystallography, a crystal is first prepared and mounted on a goniometer, which allows for precise rotation and tilting of the crystal relative to the X-ray beam. The crystal is then placed in an X-ray diffractometer, where a monochromatic X-ray beam is directed onto the crystal. As the X-rays interact with the crystal lattice, they are scattered in different directions, producing a diffraction pattern on a detector. The diffraction pattern typically consists of a series of spots corresponding to the different planes of the crystal lattice (Smyth & Martin, 2000).

To determine the crystal structure from the diffraction pattern, sophisticated mathematical techniques, such as the Fourier transform, are employed to convert the diffraction data into an electron density map. This electron density map reveals the spatial distribution of electrons within the crystal, providing a detailed picture of the atomic arrangement.

X-ray crystallography has led to numerous groundbreaking discoveries and advances in various fields. In chemistry, it has been instrumental in determining the structures of complex molecules, such as enzymes and pharmaceutical compounds. X-ray crystallography has shed light on the structures of proteins and nucleic acids in biology, revealing crucial insights into their functions and interactions. In materials science, X-ray crystallography has been used to study the atomic arrangements in crystals, helping to design and engineer new materials with desirable properties.

Despite its power and versatility, X-ray crystallography does have limitations. The technique requires the availability of high-quality single crystals, which can be challenging to obtain for some materials. Additionally, X-ray crystallography is limited to studying periodic crystals and cannot be applied to disordered or amorphous materials (Abola et al., 2000).

X-ray crystallography is a cornerstone technique in the study of crystal structures, providing invaluable information about the arrangement of atoms within crystals. Its contributions to chemistry, biology, materials science, and many other disciplines have been transformative, leading to significant discoveries and advancements. As technology advances, X-ray crystallography will undoubtedly remain a vital tool in scientific research, helping unravel the mysteries of the atomic and molecular world (Rousse et al., 2001).

8.7. NUCLEAR MAGNETIC RESONANCE (NMR) SPECTROSCOPY

Nuclear magnetic resonance (NMR) Spectroscopy is a sophisticated analytical technique widely used in chemistry, biochemistry, and related fields to study molecules' structure, dynamics, and interactions. NMR spectroscopy exploits the magnetic properties of certain atomic nuclei, such as hydrogen (1H), carbon-13 (13C), and nitrogen-15 (15N), in the presence of a strong magnetic field. This non-destructive and non-invasive technique has become indispensable for understanding molecular structures, explaining reaction mechanisms, and investigating biological processes.

The fundamental principle of NMR spectroscopy is based on the fact that atomic nuclei with an odd number of protons or neutrons possess a net nuclear spin, giving rise to a magnetic moment. When placed in a strong magnetic field, these nuclei align either with or against the magnetic field. When subjected to a radiofrequency pulse, the nuclei absorb energy and transition between spin states. The absorbed energy is then re-emitted as radiofrequency radiation when the nuclei return to their original spin states. The frequency at which this energy is absorbed and re-emitted, known as the resonance frequency, is directly proportional to the strength of the magnetic field and the properties of the surrounding molecular environment (Rehr & Kauzlarich, 1994) (Figure 8.5).

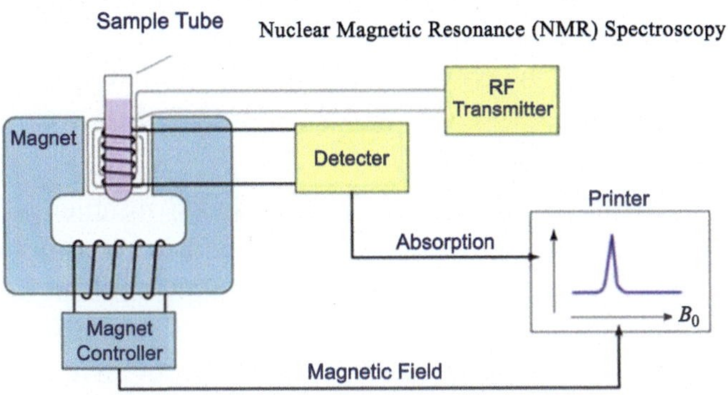

Figure 8.5. *Principle of NMR spectroscopy.*

Source: Brian, Creative Commons License.

In NMR spectroscopy, the sample is placed in a magnetic field, and a series of radiofrequency pulses are applied to excite the nuclei and create a measurable NMR signal. By varying the strength of the magnetic field or the frequency of the radiofrequency pulses, NMR spectroscopists can gather valuable information about the chemical shifts, coupling constants, and relaxation times of the nuclei within the sample.

One of the significant advantages of NMR spectroscopy is its ability to provide detailed structural information about molecules in solution. It allows researchers to determine the connectivity of atoms, the conformational flexibility of molecules, and the arrangement of functional groups within a compound. As a result, NMR spectroscopy is widely used in organic chemistry for structure determination, verification of synthesized compounds, and identification of unknown substances (Bothwell & Griffin, 2011). In biochemistry, NMR spectroscopy is crucial in studying the structure and dynamics of biomolecules such as proteins, nucleic acids, and carbohydrates. It can reveal insights into protein folding, ligand-binding interactions, and enzymatic reactions, helping unravel the intricate biological processes' intricate mechanisms.

The information obtained from NMR spectroscopy is complemented by two-dimensional (2D) and multidimensional NMR techniques. These techniques allow researchers to correlate signals from neighboring nuclei and provide more detailed structural information, particularly in complex molecules or mixtures.

Despite its powerful capabilities, NMR spectroscopy does have limitations. It requires high-quality samples, and the sensitivity of NMR signals decreases with increasing molecular size. Additionally, acquiring multidimensional NMR data can be time-consuming, especially for large biomolecules.

NMR spectroscopy is a versatile and indispensable technique for investigating molecules' structure, dynamics, and interactions in solution. Its ability to provide detailed structural information and its non-destructive nature have made it an essential tool in chemistry, biochemistry, and related fields. As technology advances, NMR spectroscopy will remain at the forefront of scientific research, driving discoveries and advancements in various disciplines (Elipe, 2003).

8.8. LASER SPECTROSCOPY

Laser Spectroscopy is a powerful and versatile technique that utilizes laser light to investigate the properties of atoms, molecules, and other particles.

This analytical method has revolutionized various scientific fields, including physics, chemistry, biology, and environmental science. Laser spectroscopy offers precise and sensitive measurements, allowing researchers to study the interactions between light and matter with exceptional accuracy and resolution.

The foundation of laser spectroscopy lies in the interaction between laser light and the particles of interest. Laser light at a sample can induce various spectroscopic processes, such as absorption, emission, scattering, and fluorescence. These processes provide valuable information about the sample's energy levels, electronic transitions, and molecular structures (Staudacher et al., 2013).

One of the key advantages of laser spectroscopy is its ability to tune the laser light to specific wavelengths, enabling selective excitation of the sample's energy levels. This tunability allows researchers to target specific transitions and probe particular atomic or molecular states with high precision (Figure 8.6).

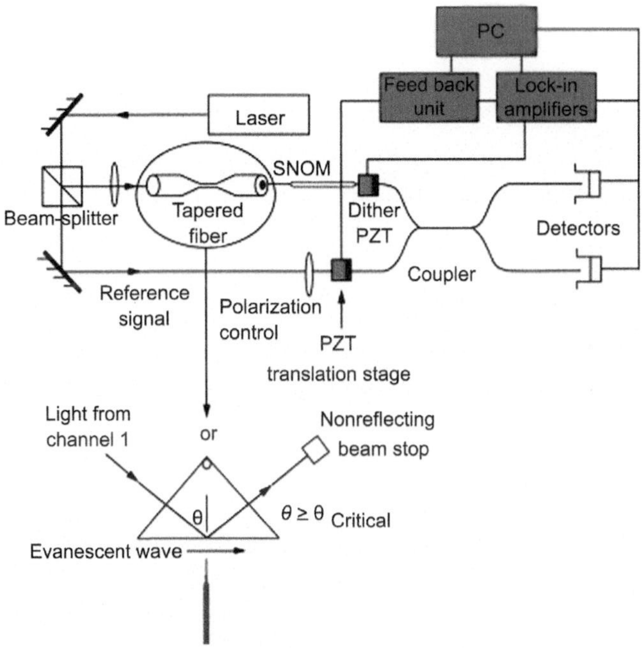

Figure 8.6. *Scheme of laser spectrometry.*

Source: P. Werle, Creative Commons License.

In laser absorption spectroscopy, the sample absorbs specific wavelengths of laser light, providing information about the particles' energy levels and electronic structure. This technique is widely used in gas-phase spectroscopy, which helps identify and quantify the concentration of molecules in a sample.

Laser-induced fluorescence spectroscopy involves exciting the sample with laser light and measuring the emitted fluorescence radiation. This technique is extensively used in biochemistry and molecular biology to study the structure and dynamics of biomolecules, such as proteins and DNA (Campbell et al., 2016).

Raman spectroscopy is another essential application of laser spectroscopy, where laser light interacts with a sample, causing a small fraction of the light to be scattered at different wavelengths. This scattered light carries information about molecules' vibrational and rotational modes, allowing researchers to identify chemical compositions and study molecular vibrations.

Laser spectroscopy has also found applications in environmental monitoring and trace gas detection. Techniques such as cavity-enhanced absorption spectroscopy (CEAS) and laser-induced breakdown spectroscopy (LIBS) enable highly sensitive and selective detection of pollutants and trace elements in the atmosphere and industrial environments (Thompson, 1985).

In addition to its analytical capabilities, laser spectroscopy plays a crucial role in fundamental physics research. For instance, laser cooling and trapping techniques have allowed scientists to achieve ultra-cold temperatures, creating Bose-Einstein condensates and groundbreaking studies of quantum phenomena.

While laser spectroscopy offers many advantages, it does have certain challenges. High-precision measurements require precise calibration and stability of the laser source. Additionally, the interactions between laser light and samples can lead to photodegradation or photodamage, particularly in biological samples (Ito et al., 1996).

Laser Spectroscopy is a versatile and powerful technique that has transformed our understanding of the interactions between light and matter. Its high precision, selectivity, and sensitivity have made it an indispensable tool in various scientific fields, from fundamental physics research to environmental monitoring and biochemistry. As technology advances, laser spectroscopy will undoubtedly remain at the forefront of scientific discoveries, driving innovations and furthering our knowledge of the atomic and molecular world (Schawlow, 1978).

8.9. TIME OF FLIGHT MASS SPECTROMETRY

Time-of-Flight Mass Spectrometry (TOF-MS) is a powerful and widely used technique for analyzing ions based on their mass-to-charge ratio (m/z). It is a versatile and high-resolution mass spectrometry method that has applications in various scientific disciplines, including chemistry, biochemistry, environmental, and materials science.

The basic principle of TOF-MS involves measuring the time ions travel from the ionization source to the detector. The process begins with the ionization of a sample, where molecules are converted into ions by various ionization techniques such as electrospray ionization (ESI), matrix-assisted laser desorption/ionization (MALDI), or electron impact ionization. An electric field accelerates The generated ions into a flight tube. The speed at which the ions travel through the flight tube is proportional to their m/z value (Mamyrin, 2001) (Figure 8.7).

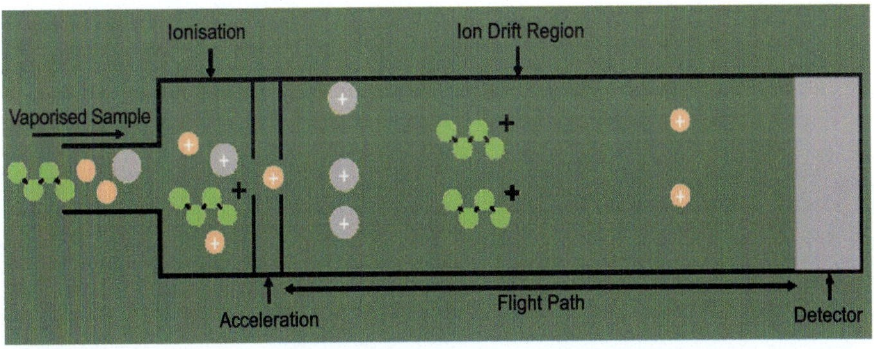

Figure 8.7. *Principle of time of flight mass spectrometry.*

Source: P. Torrione, Creative Commons License.

As the ions reach the end of the flight tube, they strike a detector, typically a microchannel plate or a time-sensitive detector, generating an electrical signal. By recording the time-of-flight for each ion, the instrument can calculate the m/z value of the ions with high accuracy and precision. The resulting mass spectrum represents the distribution of ions based on their masses, providing valuable information about the sample's molecular composition and structure.

One of the key advantages of TOF-MS is its ability to rapidly measure a wide range of masses. Because the flight time of the ions is independent of their initial kinetic energy, TOF-MS is considered a non-scanning mass spectrometry technique, allowing for high-speed data acquisition. This feature is particularly useful in applications requiring rapid complex mixture analysis (Boesl, 2017).

TOF-MS is well-suited for analyzing large biomolecules such as proteins, peptides, and nucleic acids, making it a valuable tool in proteomics and genomics research. It can identify and quantify biomolecules based on their masses, providing essential insights into biological processes, post-translational modifications, and protein-protein interactions.

In addition to its applications in the life sciences, TOF-MS is also used in environmental analysis to detect pollutants, contaminants, and trace elements. Its high sensitivity and resolution enable the precise identification and quantification of compounds in environmental samples.

Despite its advantages, TOF-MS does have some limitations. For high-resolution applications like isotopic analysis, specialized mass analyzers like quadrupole or ion trap mass spectrometers may offer better performance. Additionally, TOF-MS may have limited sensitivity for low-abundance ions compared to other mass spectrometry techniques (Chernushevich et al., 2001).

Time-of-Flight Mass Spectrometry (TOF-MS) is a versatile and powerful analytical technique for determining the masses of ions with high resolution and accuracy. Its rapid data acquisition and ability to analyze complex mixtures make it an essential tool in various scientific fields, from proteomics and genomics research to environmental analysis and materials science. As technology advances, TOF-MS will remain a critical tool in scientific research, driving discoveries and innovations across various disciplines (Cotter, 1992).

8.10. SUMMARY

This chapter explored various modern techniques used to study atoms and related phenomena in various scientific disciplines. These cutting-edge techniques have significantly advanced the understanding of the atomic and molecular world, providing unprecedented insights into atoms' structures, properties, and interactions.

The chapter began with an introduction to the fundamental nature of atoms and their crucial role in modern chemistry, illustrated the concept of the atom as the basic building block of matter, and finally discussed the conservation of matter and energy, macroscopic, and nanoscopic methods, and the physical structure of atoms. The first set of techniques examined in the chapter focused on imaging and analyzing materials' surface and internal structures at the atomic level. Scanning Electron Microscopy (SEM) was highlighted for its ability to reveal materials' surface morphology and topography. At the same time, Transmission Electron Microscopy (TEM) took center stage for its unparalleled capability to visualize the internal atomic structures of materials. AFM and STM were also discussed for their roles in investigating surface topography, mechanical properties, and manipulation of individual atoms.

Moving beyond imaging, the chapter explored techniques that provide detailed information about atomic and molecular structures. X-ray Crystallography emerged as a pivotal tool for determining the atomic and molecular structures of crystals, enabling researchers to understand the spatial arrangement of atoms within the crystal lattice. NMR spectroscopy was praised for its ability to probe atomic nuclei's electronic and spatial environments within molecules, explaining molecular structures and dynamics.

With its diverse applications, Laser Spectroscopy demonstrated its prowess in investigating the interactions between light and matter at the atomic and molecular levels. Time-of-Flight Mass Spectrometry (TOF-MS) was celebrated for its high-resolution analysis of ions based on their mass-to-charge ratio, providing valuable information about the masses and elemental compositions of atoms and molecules.

Throughout the chapter, it became evident that these modern techniques have played vital roles in advancing scientific research across various fields, including chemistry, biochemistry, physics, materials science, and environmental science. They have paved the way for groundbreaking discoveries, enabling us to explore and manipulate the atomic world with unprecedented precision and accuracy.

In conclusion, the chapter showcased a remarkable array of powerful and versatile techniques that have transformed the understanding of atoms and their significance in scientific exploration. From imaging surface structures to probing molecular dynamics, these techniques have revolutionized the approach to studying atoms, driving innovations, and contributing to scientific breakthroughs in diverse fields. As technology evolves, these modern techniques will likely remain indispensable tools in advancing the knowledge of the atomic and molecular universe.

REFERENCES

1. Abola, E., Kuhn, P., Earnest, T., & Stevens, R. C., (2000). Automation of X-ray crystallography. *Nature Structural Biology*, *7*(11), 973–977.
2. Akhtar, K., Khan, S. A., Khan, S. B., & Asiri, A. M., (2018). Scanning electron microscopy: Principle and applications in nanomaterials characterization. *Handbook of Materials Characterization*, *1*(1), 113–145.
3. Alario, M. V., & Freudenburg, W. R., (2007). Atoms for peace, atoms for war: Probing the paradoxes of modernity. *Sociological Inquiry*, *77*(2), 219–240.
4. Albanese, A., & Vicentini, M., (1997). Why do we believe that an atom is colorless? Reflections about the teaching of the particle model. *Science & Education*, *6*, 251–261.
5. Artusio-Glimpse, A., Simons, M. T., Prajapati, N., & Holloway, C. L., (2022). Modern RF measurements with hot atoms: A technology review of Rydberg atom-based radio frequency field sensors. *IEEE Microwave Magazine*, *23*(5), 44–56.
6. Binnig, G., & Rohrer, H., (1983). Scanning tunneling microscopy. *Surface Science*, *126*(1–3), 236–244.
7. Binnig, G., & Rohrer, H., (1985). The scanning tunneling microscope. *Scientific American*, *253*(2), 50–58.
8. Binnig, G., & Rohrer, H., (2000). Scanning tunneling microscopy. *IBM Journal of Research and Development*, *44*(1/2), 279.
9. Binnig, G., Rohrer, H., Gerber, C., & Weibel, E., (1982). Surface studies by scanning tunneling microscopy. *Physical Review Letters*, *49*(1), 57.
10. Boesl, U., (2017). Time - of - flight mass spectrometry: introduction to the basics. *Mass Spectrometry Reviews*, *36*(1), 86–109.
11. Bothwell, J. H., & Griffin, J. L., (2011). An introduction to biological nuclear magnetic resonance spectroscopy. *Biological Reviews*, *86*(2), 493–510.
12. Bulienkov, N. A., (2011). Systemic structural modular generalization of the crystallography of bound water applied to study the mechanisms of processes in biosystems at the atomic and molecular level. *Crystallography Reports*, *56*, 680–697.
13. Campbell, P., Moore, I. D., & Pearson, M. R., (2016). Laser spectroscopy for nuclear structure physics. *Progress in Particle and Nuclear Physics*, *86*, 127–180.

14. Cappella, B., & Dietler, G., (1999). Force-distance curves by atomic force microscopy. *Surface Science Reports*, *34*(1–3), 1–104.
15. Chernushevich, I. V., Loboda, A. V., & Thomson, B. A., (2001). An introduction to quadrupole–time - of - flight mass spectrometry. *Journal of Mass Spectrometry*, *36*(8), 849–865.
16. Cotter, R. J., (1992). Time-of-flight mass spectrometry for the structural analysis of biological molecules. *Analytical Chemistry*, *64*(21), 1027A–1039A.
17. Del Río, E., (1996). The body as foundation of the screen: Allegories of technology in atom Egoyan's speaking parts. *Camera Obscura-Berkeley Then Los Angeles Then Bloomington-*, 93–116.
18. Duke, C. B., (1984). Atoms and electrons at surfaces: A modern scientific revolution. *Journal of Vacuum Science & Technology A: Vacuum, Surfaces, and Films*, *2*(2), 139–143.
19. Elipe, M. V. S., (2003). Advantages and disadvantages of nuclear magnetic resonance spectroscopy as a hyphenated technique. *Analytica Chimica Acta*, *497*(1–2), 1–25.
20. Garcıa, R., & Perez, R., (2002). Dynamic atomic force microscopy methods. *Surface Science Reports*, *47*(6–8), 197–301.
21. Giessibl, F. J., (2003). Advances in atomic force microscopy. *Reviews of Modern Physics*, *75*(3), 949.
22. Goldman, N., Juzeliūnas, G., Öhberg, P., & Spielman, I. B., (2014). Light-induced gauge fields for ultracold atoms. *Reports on Progress in Physics*, *77*(12), 126401.
23. Grafström, S., (2002). Photoassisted scanning tunneling microscopy. *Journal of Applied Physics*, *91*(4), 1717–1753.
24. Guilhaus, M., (1995). Special feature: Tutorial. Principles and instrumentation in time - of - flight mass spectrometry. Physical and instrumental concepts. *Journal of Mass Spectrometry*, *30*(11), 1519–1532.
25. Hansma, P. K., & Tersoff, J., (1987). Scanning tunneling microscopy. *Journal of Applied Physics*, *61*(2), R1–R24.
26. Hansma, P. K., Elings, V. B., Marti, O., & Bracker, C. E., (1988). Scanning tunneling microscopy and atomic force microscopy: Application to biology and technology. *Science*, *242*(4876), 209–216.
27. Inkson, B. J., (2016). Scanning electron microscopy (SEM) and transmission electron microscopy (TEM) for materials characterization.

In: *Materials Characterization Using Nondestructive Evaluation (NDE) Methods 1*(1), pp. 17–43.

28. Ito, H., Nakata, T., Sakaki, K., Ohtsu, M., Lee, K. I., & Jhe, W., (1996). Laser spectroscopy of atoms guided by evanescent waves in micron-sized hollow optical fibers. *Physical Review Letters, 76*(24), 4500.
29. Kohl, H., & Reimer, L., (2008). Transmission electron microscopy. *Springer Series in Optical Sciences, 1*(1), *36*.
30. Kuk, Y., & Silverman, P. J., (1989). Scanning tunneling microscope instrumentation. *Review of Scientific Instruments, 60*(2), 165–180.
31. Kuznetsova, T. G., Starodubtseva, M. N., Yegorenkov, N. I., Chizhik, S. A., & Zhdanov, R. I., (2007). Atomic force microscopy probing of cell elasticity. *Micron, 38*(8), 824–833.
32. Leamy, H. J., (1982). Charge collection scanning electron microscopy. *Journal of Applied Physics, 53*(6), R51–R80.
33. Mamyrin, B. A., (1994). Laser assisted reflectron time-of-flight mass spectrometry. *International Journal of Mass Spectrometry and Ion Processes, 131*, 1–19.
34. Mamyrin, B. A., (2001). Time-of-flight mass spectrometry (concepts, achievements, and prospects). *International Journal of Mass Spectrometry, 206*(3), 251–266.
35. Meyer, E. R. N. S. T., (1992). Atomic force microscopy. *Progress in Surface Science, 41*(1), 3–49.
36. Mohammed, A., & Abdullah, A., (2018). Scanning electron microscopy (SEM): A review. In: *Proceedings of the 2018 International Conference on Hydraulics and Pneumatics—HERVEX, B ile Govora, Romania, Vol. 2018*, pp. 7–9.
37. Nandi, T., & Ainavarapu, S. R. K., (2021). Applications of atomic force microscopy in modern biology. *Emerging Topics in Life Sciences, 5*(1), 103–111.
38. Newbury, D. E., & Williams, D. B., (2000). The electron microscope: the materials characterization tool of the millennium. *Acta Materialia, 48*(1), 323–346.
39. Nshii, C. C., Vangeleyn, M., Cotter, J. P., Griffin, P. F., Hinds, E. A., Ironside, C. N., & Arnold, A. S., (2013). A surface-patterned chip as a strong source of ultracold atoms for quantum technologies. *Nature Nanotechnology, 8*(5), 321–324.

40. Pla, J. J., Tan, K. Y., Dehollain, J. P., Lim, W. H., Morton, J. J., Jamieson, D. N., & Morello, A., (2012). A single-atom electron spin qubit in silicon. *Nature*, *489*(7417), 541–545.
41. Rehr, A., & Kauzlarich, S. M., (1994). From international tables for X-ray crystallography. *Acta Cryst*, *50*, 1859–1861.
42. Rousse, A., Rischel, C., & Gauthier, J. C., (2001). Femtosecond X-ray crystallography. *Reviews of Modern Physics*, *73*(1), 17.
43. Rugar, D., & Hansma, P., (1990). Atomic force microscopy. *Physics Today*, *43*(10), 23–30.
44. Schawlow, A. L., (1978). Laser spectroscopy of atoms and molecules: The abundance of new laser techniques is making possible a variety of spectroscopic experiments. *Science*, *202*(4364), 141–147.
45. Scheinfein, M. R., Unguris, J., Kelley, M. H., Pierce, D. T., & Celotta, R. J., (1990). Scanning electron microscopy with polarization analysis (SEMPA). *Review of Scientific Instruments*, *61*(10), 2501–2527.
46. Schorb, M., Haberbosch, I., Hagen, W. J., Schwab, Y., & Mastronarde, D. N., (2019). Software tools for automated transmission electron microscopy. *Nature Methods*, *16*(6), 471–477.
47. Smith, D. J., (2015). Characterization of nanomaterials using transmission electron microscopy. *Nanocharacterisation*, *37*, 1–29.
48. Smyth, M. S., & Martin, J. H. J., (2000). X-ray crystallography. *Molecular Pathology*, *53*(1), 8.
49. Somorjai, G. A., (1996). Modern surface science and surface technologies: An introduction. *Chemical Reviews*, *96*(4), 1223–1236.
50. Spurgeon, S. R., Ophus, C., Jones, L., Petford-Long, A., Kalinin, S. V., Olszta, M. J., & Taheri, M. L., (2021). Towards data-driven next-generation transmission electron microscopy. *Nature Materials*, *20*(3), 274–279.
51. Staudacher, T., Shi, F., Pezzagna, S., Meijer, J., Du, J., Meriles, C. A., & Wrachtrup, J., (2013). Nuclear magnetic resonance spectroscopy on a (5-nanometer) 3 sample volume. *Science*, *339*(6119), 561–563.
52. Tanner, G., Richter, K., & Rost, J. M., (2000). The theory of two-electron atoms: between ground state and complete fragmentation. *Reviews of Modern Physics*, *72*(2), 497.
53. Tersoff, J., & Hamann, D. R., (1983). Theory and application for the scanning tunneling microscope. *Physical Review Letters*, *50*(25), 1998.

54. Tersoff, J., & Hamann, D. R., (1985). Theory of the scanning tunneling microscope. *Physical Review B*, *31*(2), 805.
55. Thompson, R. C., (1985). High resolution laser spectroscopy of atomic systems. *Reports on Progress in Physics*, *48*(4), 531.
56. Urban, K. W., (2008). Studying atomic structures by aberration-corrected transmission electron microscopy. *Science*, *321*(5888), 506–510.
57. Vernon-Parry, K. D., (2000). Scanning electron microscopy: An introduction. *III-Vs Review*, *13*(4), 40–44.
58. Winey, M., Meehl, J. B., O'Toole, E. T., & Giddings Jr, T. H., (2014). Conventional transmission electron microscopy. *Molecular Biology of the Cell*, *25*(3), 319–323.
59. Zhou, W., Apkarian, R., Wang, Z. L., & Joy, D., (2007). Fundamentals of scanning electron microscopy (SEM). *Scanning Microscopy for Nanotechnology: Techniques and Applications*, *1*(1), 1–40.

INDEX